Performance, Transparency, and the Cultures of Surveillance

Performance, Transparency, and the Cultures of Surveillance

James M. Harding

UNIVERSITY OF MICHIGAN PRESS
ANN ARBOR

Published in the United States of America by
the University of Michigan Press
Manufactured in the United States of America
⊗ Printed on acid-free paper

2021 2020 2019 2018 4 3 2 1

A CIP catalog record for this book is available from the British Library.

ISBN 978-0-472-13074-0 (hardcover : alk. paper)
ISBN 978-0-472-03709-4 (paper : alk. paper)
ISBN 978-0-472-12363-6 (e-book)

Acknowledgments

For theater historians and performance studies scholars, for political activists and organizers, for anyone interested in community, the collective, or the commons, the acknowledgments page of a book ought to provide a moment of reassurance. Easily dismissed as a mere formality, this small preface found in the opening pages of almost every scholarly work is in fact a reminder—one that cannot be overrehearsed—that the production of knowledge is always a collaborative affair. And isn't this what the term "to acknowledge" means with its simultaneous reference to knowledge itself, to a recognition of receipt, and to an expression of indebtedness and gratitude? With regard to this book, which at some level is a book like any other, I do not want to lose sight of the extent to which any knowledge it seems to profess needs to be embedded within an unequivocal act of acknowledgment that, beyond being a mere formality, is in fact a call to rethink what we imagine knowledge to be. For I cannot help but wonder how much different our sense of epistemology might be—how much more radical our theories of knowledge could be—if only we were to conceptualize knowledge not as something we possess but rather as something that we receive on loan and as something that is akin to a debt whose reward somehow always exceeds what one must do to repay it. At some point, we must realize—or at the very least, I must admit what others perhaps already recognize—that knowledge is not so much something we choose to share as it is something that by its very nature is shared. We can speak of common knowledge as something superficially shared, but it behooves us recognize critical knowledge as a product, indeed as a fundamental characteristic, of the commons. With such shared critical knowledge we build community, and what this book represents is the workings of a critical community from which I have benefited greatly.

That community includes friends and colleagues both old and new, near and far, invited and unexpected. It includes colleagues at two different universities on two different continents where I have had the privilege of teaching. Serious work on this book began shortly after I joined the faculty in the School of Theatre, Performance and Cultural Policy Studies at the University of Warwick in the United Kingdom. I owe a very special note of thanks there to Tim White for his unbelievable generosity in sharing a wealth of material with me, to Wallace McDowell and Susan Haedicke for absolutely inspiring professional and personal friendships, to Jim Davis, Janelle Reinelt and Baz Kershaw for sharing their vision and insight, to Saul Hewish for (discussing) his prison theater activism and to Nadine Holdsworth for her steady leadership as Head of School. I am equally grateful to Nicolas Whybrow, Yvette Hutchison, Milija Gluhovic, and Silvija Jestrovic for invaluable conversations during my time at Warwick.

Across the ocean at the University of Maryland, where I finished the final chapters of this book, I have also been fortunate enough to have colleagues who are no less generous in sharing their knowledge and insights. In the School of Theatre, Dance and Performance Studies at Maryland, I am particularly grateful to my immediate colleagues in the graduate program in Theatre and Performance Studies. Since the governing assumptions of this book are aligned with the field of performance studies, I am especially grateful to my colleague Laurie Frederik, who has been an important sounding-board for many of the theoretical arguments in the latter half of this book. Faedra Carpenter and Esther Kim Lee have been no less gracious in their willingness to lend a critical ear to the arguments that ultimately found their way into these pages. And to Frank Hildy, I am grateful for countless bits of historical insight. In the larger community of the School of Theatre, Dance and Performance Studies, I have also benefited greatly, first from conversations with my colleagues Scot Reese, Mitchell Hébert, Karen Bradley, Dan Conway, Brian MacDevitt, Jared Mezzocchi, Misha Kachman, and Jennifer Barclay and second from conversations with our graduate students, especially Christina Banalopoulou, Jeff Kaplan, Les Gray, and Jonelle Walker. Special thanks also go to Victoria Scrimer for her work on the book's index. I am also indebted to colleagues across the campus: to Jason Farman in American Studies, to Hashan Elahi in the Department of Art, and Orrin Wang and Julius Fleming in the Department of English. But above all, the completion of this book would not have been possible without the unbelievable support that Leigh Smiley has extended to me in her role as Director of the School of Theatre, Dance and Performance Studies. I am grateful to her for her leadership, her collegiality, and her friendship.

Looking beyond the institutions where I have taught, I know that this book is the result of countless exchanges with colleagues in the profession at conferences, symposiums, and lectures. Here the names are many, but three stand out in particular: Sarah Bay-Cheng, Elise Morrison, and Markus Wessendorf. In no uncertain terms, Sarah and Elise have been my most important interlocutors for this project, not only on the panels that we have organized or where we have simply presented together but also in the invaluable conversations and discussions that I have had with them between sessions at conferences or over email. With Sarah many of these exchanges were a continuation of ongoing dialogues that we have had because of a shared interest in avant-garde performance. But I have benefited immensely from the recent professional friendship that has emerged with Elise as a result of our parallel scholarly projects on surveillance and performance. I am grateful to her for publishing a shorter version of my chapter on the Surveillance Camera Players in the special issue of the *International Journal of Performance Arts and Digital Media* that she edited in 2015, and I hope that my book meets the high standard that she has already set with her marvelous book *Discipline and Desire: Surveillance Technologies in Performance*. The more unexpected friendship to come from this project is the one I now enjoy with Markus Wessendorf, whose invitation to give a keynote address at a symposium on surveillance and performance that he organized at the Goethe-Universität in Frankfurt, Germany, in 2015 provided the foundation for what became the first chapter of this book. He also introduced me to Jan-Christoph Gockel, Thomas Halle, and Konstantin Küspert's theater piece *Ich Bereue Nichts* as well as to the performance work of Christoph Faulhaber. Discussions of both figure prominently in the book's final chapter. My collaborations with Marcus have continued in Hawaii and in Hamburg. I am immensely grateful to him for the invitations he has extended to me, the general support he has provided, and an understanding of the nuances of the cultures of surveillance that I have found invaluable in completing this book.

Similar notes of gratitude need to be extended to Jennifer Parker-

Starbuck and Josh Abrams for an invitation to present my research at the London Theatre Seminar as well as to those who extended invitations for other lectures: to Kirsten Shepherd-Barr at Oxford; to Patrick Duggin and Kara Reilly at the University of Exeter; to Kati Röttger and Sruti Bali at the University of Amsterdam; to John Rouse at the University of California, San Diego; to Peter Eckersall and Frank Hentschker at the CUNY Grad Center and the Martin E. Segal Theater Center; to Miriam Felton-Dansky and Jacob Gallagher-Ross at Bard College and the University of Toronto; to Erika Fischer-Lichte and Christel Weiler at the International Research Center / Interweaving Performance Cultures at the Freie Universität, Berlin; and last but not least, to Steve Wilmer at Trinity College, Dublin, who also published an earlier version of my chapter on dataveillance and Manjula Padmanabhan's *Harvest* in his edited anthology *Resisting Biopolitics: Philosophical, Political and Performative Strategies* (Routledge, 2015).

Parallel to writing this book, I was also editing the anthology, *The Sixties, Center Stage* (Michigan, 2017) with my longtime collaborator and close friend, Cindy Rosenthal. In practical terms, our collaboration on that anthology meant that, amid our conversations about the anthology, I also talked to her about almost every part of this book. I am indebted to her for sound advice and for frank criticism. In a similar respect, this book has benefited greatly from the ongoing dialogues and insightful conversations about theater, performance, and politics that I continue to have with friends like Mike Sell, Jean Graham-Jones, Kimberly Jannarone, Kate Bredeson, Michael Chemers, Andrew Sofer, Adam Versényi, Marc Silberman, Jim Ball, David Savran, Maurya Wickstrom, Nicholas Ridout, Paul Rae, and Stephen Bottoms (to whom I am especially grateful for his willingness to share material on the Stanford Prison Experiment).

I want to offer a special note of thanks—the kind of thanks that I have expressed many times in the past—to LeAnn Fields at the University of Michigan Press. Her support of my work, which extends back to the 1990s and our initial conversations about the anthology *Contours of the Theatrical Avant-Garde* (Michigan, 2000), has been nothing short of decisive in my career. I cannot imagine this present book or almost any of my previous books or anthologies without the wise, supportive counsel that she has given me over the years. She has believed in my scholarship, and she has cultivated it. I am delighted to call her a friend, but I am even more honored to have her as an editor. I am grateful to her for the work that she, Marcia LaBrenz, and the production staff at the University of Michigan Press have put into this book.

Professional and scholarly relationships aside, this book has come at no small cost to my family. Here it is not enough to thank my partner, Friederike Eigler, or our two sons, Lukas and Daniel Eigler-Harding, for their support. This project and the career path that has accompanied it have often demanded sacrifices that were not of their own choosing. It is one thing to say that the writing of it occurred on two different continents at two different universities; it is another to acknowledge the strain that my commuting back and forth between the United Kingdom and the United States and that my writing have meant for my partner, who has her own scholarship and career obligations as a Professor of German Literature at Georgetown University; for our youngest son, Daniel, who was in in first years of high school while I was a member of the faculty at Warwick and commuting back and forth across the Atlantic; or for our oldest son, who was then in his first years at the Rhode Island School of Design. They have given much to this book—sometimes in conversations with me about it but more often simply in the time and space I have needed to write. In the wider circle of my family, I am also indebted to my father-in-law, Friedrich Wilhelm Eigler, and to my brother-in-law, Martin Eigler, for important conversations about key aspects of this project. Returning to the immediate family of my wife and our two sons, I hope that in some small way this short note touches upon the profound debt that I owe to each of them, especially to Friederike. They sustain me.

The final note in my acknowledgments belongs, first, to my father, Charles F. Harding, who did not live to see its completion and thus never knew that I intended to dedicate the book to him and to my mother. Our best conversations had only begun. The final note also belongs to my mother, Jacqueline Harding, who passed as the book was going to press and who also did not know about the coming dedication. I miss them both.

To my father, Charles F. Harding,
and to simple acts of restoration
To my mother, Jacqueline Harding,
and to the struggle against silence

Contents

All Eyes to the Street: Surveillance, Information Technologies, and the Space of Brecht's "Street Scene"

Listening to the Warm-Up Band: A Preface to an Introduction

Let us begin by reprising a narrative that I heard the dissident (former) East German folk singer Wolf Biermann reprise at a small concert at Georgetown University in November 1992, just three years after the fall of Berlin wall—a concert that was itself a kind of nostalgic reprisal of Biermann's greatest hits. Biermann is perhaps best known for having been stripped of his East German citizenship in 1976 while he was on an officially sanctioned concert tour in West Germany, where he had a popular following among students and activists on the left. His visa to give a concert in the West had come as a surprise since as a lyricist Biermann has always had an unflinching and biting sense of political satire that is quite reminiscent of Brecht and Weill and that was unsparingly critical of the East German authorities. But by the mid-1970s he was internationally famous and thus a bit of a thorny problem. So prior to revoking his citizenship, the Stasi (the East German security police) had subjected Biermann to extensive surveillance, assigning at various times seventy to two hundred agents to observe him.[1] As Hollywood has encouraged us all to expect, the Stasi also bugged Biermann's home, to which Biermann responded by penning what became one of his most famous songs: "Die Stasi Ballade," or "The Stasi Ballad." This song is written and sung as a first-person address and it speaks directly to the agents that Biermann rightly assumed were meticulously recording every word spoken in his apartment. As Amy Schwartz has noted, the song "hailed the invisible audience of agents" as Bierman's "truest fans," who will "ensure . . . [his] immortality" as "they pore over the drudgery of transcription, spool the tapes repeatedly to get the exact wording, and, inevitably, go to bed humming his tunes."[2]

Biermann told his Georgetown University audience that after the fall of the wall, he gained access to his Stasi files. At the building of the federal commissioner for the Stasi records, he was charmed, he said, to discover that in fact he had been right all along, since he chanced upon a transcribed version of the "Stasi Ballad" among the different reports in his files. But reading through the transcription, he was also surprised to discover one glaring error. In the original, the chorus written by Biermann reads "Die Stasi ist mein Eckermann" ("the Stasi is my Eckermann"). It thus amusingly equates the Stasi to Johann Peter Eckermann, who was the nineteenth-century literary secretary of Johann Wolfgang von Goethe.[3] But in the entry to Biermann's Stasi file where he found the transcription of his song, the agent had written something different. The chorus reads instead: "Die Stasi ist mein Henkerman" ("the Stasi is my executioner"). Before his Georgetown audience, Biermann speculated that the agent didn't understand the original literary reference, and, fearing a reprimand from his or her superior for an incomplete transcription, the agent came up with an original rhyme. Many years later, all of this made for a great story, especially since Biermann concluded it with a surprise concession. Much to everyone's delight, he told the audience that after comparing his original line with the Stasi agent's improvised alternative, it was clear that the agent was a better poet.

Entertaining though this anecdote might be, I want to suggest as the point of departure for this book that we bracket the humor momentarily, that we reject the real possibility that the agent simply misheard, and that we take Biermann's concluding assessment at face value. It is worth doing so if only because Biermann's narrative sets the stage for the dialogue that this book orchestrates between performance and surveillance as well as between the disciplines of performance studies and surveillance studies. Like the surreal image of Biermann singing from his apartment to the Stasi agent improvising his or her transcription in the adjacent room—above, below, or maybe next door?—this book is as much about how surveillance complicates performance as it is about how the concept of performance might provide a more nuanced understanding of surveillance as a sociopolitical phenomenon. In this respect, Biermann's subsequent willingness to measure his skills as lyricist against those of his Stasi counterpart is not so much about abstract poetic skills as it is about how different subject positions within surveillance regimes elicit qualitatively different kinds of performances: empowering some, disarming some, and dictating others.

We might shift the focus of Biermann's narrative just enough, for example, to ask not who was the better poet but rather whose act was the more subversive: the dissident folk singer who in penning yet another song of political satire performed as expected or the Stasi agent who apparently falsified records and thus corrupted the integrity of the surveillance process in an improvised performance all his or her own? Such questions may not have clear answers, but they do place the concept of performance at the center of sociopolitical debates about how the techniques and the technologies of surveillance provoke specific kinds of acts that construct and regulate individual and collective identities. Such are the questions I pursue throughout this book. All of this is to say that the book is concerned with surveillance's performative acts:

Those that are integral to surveillance and that might be identified with the question, how does surveillance itself perform?

Those that surveillance promotes, mediates, or impedes and that might be identified with the question, what kinds of performance does surveillance produce?

And finally, those that seek to resist or counter surveillance in its myriad forms and that might be identified with the question, what performances disrupt surveillance and its assemblages?

With regard to this latter question in particular, the poetic excursions of the Stasi agent who filed the false report on Biermann's "Stasi Ballad" are particularly interesting because, whether we realize it or not, as surveillance proliferates and smart technologies become increasingly embedded in our daily lives, we become more and more like that agent (filing thousands of reports in the form of automatic digital updates) and Biermann becomes more and more of an anachronism. The small anecdote about the Stasi agent thus tells us not only a lot about how late twentieth-century surveillance performed but also how it was disrupted, and the quaintness of Biermann's narrative is a stark reminder that if the GDR had had today's surveillance technologies at its disposal, the Berlin wall would likely still be standing.

As the case of that Stasi agent illustrates, none of these questions lend themselves to considerations that avoid the others, and so at any given moment in the book, readers are likely to find that the above-mentioned questions are woven together through the different strands of the individual chapters. My hope is that such interweaving avoids the lure of an overly reductive binary between surveillance societies and performance communities: between those, on the one hand, who employ the technologies of surveillance in an effort to subordinate individuals to socially constructed identities that ultimately sustain entrenched and increasingly postdemocratic political orders, and those, on the other hand, who in their collective/collaborative endeavors as performance practitioners strive toward a sense of community that might effectively counter the interpellations and social sorting that surveillance technologies facilitate. As appealing as such a dichotomy might sound, the performative realities of surveillance are a lot messier, and are seldom so neatly divided. Nor are they as neatly divided as they seem to be in the scene behind Biermann's "Stasi Ballad," where he as the lone artist is defiantly positioned in opposition to a vast surveillance apparatus poised against him. Even that image is a lot messier than it seems to be.

Biermann may tell the agents assigned to his case that they will go to bed humming his songs, but they are clearly inside his head since he too will go to bed humming his song about them. And this begs the question of who among those seventy to two hundred agents assigned to his case are the friends he thinks he can trust but who, like the characters from Arthur Miller's *The Archbishop's Ceiling* whom I discuss in the opening chapter of this book, have been seduced or coerced into informing on their neighbor, their colleague, their friend, their lover, or their family member—friends like the East German writer Sascha Anderson, who in the early 1970s was imprisoned for distributing pamphlets that included lyrics form Biermann's songs and who later became the leading figure in the countercultural scene of East Berlin's Prenzlauer Berg.

After the fall of the wall in 1989 and the opening of the Stasi files,

Biermann was among those who bitterly identified his former friend Anderson (whom he publicly renamed "Sascha Asshole") as having been a Stasi informant for over fifteen years. The messy reverberations of Anderson's outing as an informant rightly led critics like Friederike Eigler to "challenge the idealized image" of Prenzlauer Berg's countercultural scene and to raise the uncomfortable question of "the counter-culture's complicity with the power structures it claimed to oppose."[4] Of course, anecdotes like Biermann's denunciation of Anderson and dramas like Miller's *The Archbishop's Ceiling* frame such reflections as Cold War narratives, but as I point out in great detail in the second chapter of this book, which deals with undercover informants among political activists, the United States and the United Kingdom have their own sordid and messy histories of undemocratic surveillance activities conducted in the name and ostensible defense of democracy itself.

Program Notes: An Overview of a Book about Surveillance and the Loss of Oversight

I have attempted to keep that messiness in focus by positioning the book's inquiries along surveillance's most problematic fault lines and contradictions, like those exemplified, for instance, in the fact that surveillance, which literally means to view from above, consistently inclines toward configurations that circumvent "oversight," or in the fact that such circumventions tend to facilitate the most egregious invasions of privacy when they result from cultivating surveillance as a private enterprise. I cite these two fault lines in particular because the book is situated where the two intersect. The book is structured conceptually around different ocular models, all of which frame a central argument that the vast privatization of surveillance technologies marks the collapse of democratic oversight and transparency. Using a wide range of examples drawn from theater and performance studies, I contend that rather than ushering in an era of increasing transparency—which is a basic paradigm of egalitarian democracies—surveillance technologies and the cultures that accompany them have in fact cultivated radically new and postdemocratic formations of power and authority that are shrouded in fortified opacity.

The first chapter of the book maps out one of the key historical backdrops to the development of those formations, and as its point of departure it takes up Karl Marx's admonishment that the critique of religion is the beginning of all critiques. In doing so, the chapter's goal is to challenge the enduring legacy of the Judeo-Christian tradition in the dominant ideologies that gave rise to and that still frame our conceptual understanding of surveillance as a sociopolitical phenomenon. Coming under the heading of "God's Eyes," the chapter uses critical studies of works like Muriel Spark's *The Abbess of Crewe*, Arthur Miller's *The Archbishop's Ceiling*, and John Patrick Shanley's *Doubt* to trace the theological and discursive foundations of that legacy in the conceptualization and justification of surveillance in Western society. In this respect, the chapter is very much about the concept of "oversight" and about the long problematic history not just of conceptualizing society as functioning beneath God's presumably benevolent and "watchful" eye, but also of co-opting this image as the conceptual frame for understanding the value of surveillance in society. The reference in the chapter's subtitle to "defrocking the cultures of surveillance" thus alludes to a critical deconstruction of surveillance's theological foundations, which is, I argue, a prerequisite for the critique of surveillance more generally.

Having stepped away then from the conceptual baggage of religious discourse, the second chapter of the book plants its feet firmly on the ground amid the political turbulence of the 1960s. It examines surveillance's excesses through the lens of undercover police operations like the FBI's COINTELPRO, which in its pursuit of what it suggested was illegal activity targeted countless legitimate political activists, often with devastating consequences. In doing so, COINTELPRO set a precedent that I trace into present-day undercover surveillance operations and that I argue is characteristic of surveillance's "shotgun approach" to the gathering of information more generally. Coming under the heading of "Private Eyes," the chapter ultimately focuses on the broad invasions of privacy that are ostensibly the "unintended consequences" of what is otherwise justified as legitimate targeted surveillance. In this respect, the chapter sustains an ongoing dialogue with Gary T. Marx's now classic study *Undercover: Police Surveillance in America* (1988), and while that dialogue carries the chapter into two deeply disturbing case studies of undercover operations against contemporary political activists in the United States, the United Kingdom, and Europe, the chapter's primary concern is with the ways that undercover surveillance highlights the performative foundations of surveillance as such. In this respect, my interest in earlier programs like COINTELPRO results from a questioning of how such programs were positioned with regard to our existing notions of theater and performance. Playing off of the sixties' political slogan that "the only theater is in the streets," the chapter argues that undercover operatives—who literally were agent provocateurs—played an instrumental role in what cultural critics have celebrated as the sixties' radical redefinition of theater and are thus every bit as vital to the broadening of our understanding of performance as wildly creative activists like Jerry Rubin or Abbie Hoffman. Not only does this awkward realization highlight a crucial performative aspect of surveillance within society, but it also gives surveillance a commanding role in the emerging notions of experimental theater and performance that now have wide scholarly currency in our profession.

The third chapter of the book moves from the low-level surveillance of undercover operatives to the maximum surveillance environments of prisons. It builds upon the arguments in the previous chapter about the performative foundations of surveillance by turning attention to the kinds of performances that surveillance generates. Coming under the heading of "Blind Eyes," the chapter considers the paradigmatic significance of Guantánamo's Camp X-Ray as a site of contemporary surveillance, and it posits it as an alternative to the eighteenth-century model of the panopticon proposed by Jeremy Bentham. The chapter's heading alludes to the fact that, unlike Bentham's "panoptic" model, Camp X-Ray is the epitome of what, in a substantial reworking of a term first coined by Didier Bigo, I call the "ban-optic," that is, a dynamic in contemporary surveillance regimes that inclines toward the maximum amount of surveillance with the least amount of oversight. Drawing upon substantial discussions of Victoria Brittain and Gillian Slovo's *Guantanamo: Honor Bound to Defend Freedom* and Philip Zimbardo's infamous Stanford Prison Experiment, the chapter argues that the island prison is thus an apt metaphor for the ways that surveillance regimes always create their own shadow, where those who have fallen under surveillance's purview are subject to violations that, while not officially sanctioned, are nonetheless tolerated as an extralegal means of asserting political authority and control. This same dynamic, I argue, is the operative consequence of neoliberalism's vast privatization of surveillance and security services and of prisons in particular. It is at the core, I argue, of how the prison industrial complex performs. It also determines which communities those performances target most directly, and here the chapter takes its most significant turn. Returning to the metaphor of the island prison—only this time to the island prison represented in Athol Fugard, John Kani, and Winston Ntshona's drama *The Island*—the chapter draws attention to the definitive role that race plays in the ban-optic performances of surveillance and, by extension, to the decisive role that the racist underpinnings of the prison industrial complex play in how the United States is constituted as a surveillance society.

Having established the performative foundations of surveillance as a sociopolitical phenomenon and having considered the models of performance that one might glean from a careful consideration of the maximum surveillance environments of prisons, chapter 4 turns consideration to the competing modes of performance that come into play when arts practitioners begin to take up the challenge of countering the expansion of surveillance technologies ever deeper into daily life. The reference in the chapter's heading to "Electronic Eyes" alludes to the most common and literally the most visible symbol of the surveillance society, the CCTV (or surveillance) camera, and the chapter itself focuses on the pathbreaking work of the Surveillance Camera Players, who started performing in front of surveillance cameras on New York City streets in 1996 and continued until they essentially disbanded in 2006 to pursue individual projects. The generative importance of this early antisurveillance activist group is only matched, I argue, by the significance of their tacit decision to disband. The chapter explores the contexts of that decision, placing it against the backdrop of the rapid advance of surveillance technologies. In many respects, the corpus of the Surveillance Camera Players' work represents an expanded variation of the kind of antisurveillance art exemplified in Wolf Biermann's "Stasi Ballad." For just as Biermann performed to the Stasi agent at the other end of the bug in his room, the Surveillance Camera Players performed in front of CCTV cameras to the officers they presumed were monitoring the screens at the other end of the cameras. But this assumption, I argue, quickly became outdated, and with the advent of digital surveillance technologies, the modes of performance refined by the Surveillance Camera

Players were literally outperformed and eclipsed by the very technologies they sought to challenge, and their performances played to an audience that had given way to computer automation.

The transition from chapter 4 to chapter 5 provides a stark reminder that the eclipse of the Surveillance Camera Players was not merely the result of the advent of a new generation of surveillance camera technologies. It was the result of surveillance technologies that rely less and less on visual formats for exerting political power and control. In this respect, the fifth chapter's heading, "Downcast Eyes," refers to a shift from the watchful eye of the surveillance camera to the expansive systems of data collection and big-data processing that have increasingly become the central mechanism of the surveillance society. In brief, chapter 5 addresses the question of how dataveillance performs, and in answering that question it looks closely at the emergence of what surveillance studies scholars have called our "data doubles," that is, the profiles of us that are compiled in data banks from the data generated by any number of our daily activities—profiles that seem to be a reflection of ourselves, that have a profound effect on what privileges we may or may not enjoy, but that we do not own. Framing these considerations within a study of dataveillance and performance in Indian playwright Manjula Padmanabhan's 1997 futuristic play *Harvest*, chapter 5 thus pays particular attention to how the emergence of data doubles has radically altered the structure of political power on both the individual and the collective levels. Specifically, the chapter argues that Padmanabhan offered an early allegorical glimpse at how dataveillance has altered the political dynamics of transparency within democratic societies not only because it is the most powerful and quickly evolving form of surveillance but because it resides almost entirely within the hands of private companies and corporations, which, as a consequence, now yield an unprecedented form of power that continues to increase as data banks grow and as the possibilities for processing their data become more and more sophisticated.

The sixth and final chapter of the book locates the rise of the surveillance society within the same historical currents that have empowered neoliberalism and that have carried us into an era that Colin Crouch has called "postdemocracy." Indeed, the chapter argues that surveillance provides postdemocracy's final fortification, and this in turn establishes the context for the chapter's pivotal question: what does the advent of postdemocracy demand of theater and performance practitioners in the twenty-first century? In asking this question, the final chapter takes stock of the cultural-political landscape that the previous chapters have set in critical relief. Its heading, "An Eye for an Eye," with its playful allusion back to the Judeo-Christian discursive traditions that were the focus of the first chapter's critique, is at some level a call to arms. It points toward the militant prescription that the chapter ultimately offers in its vision of what I call "performance postdemocracy."

That militant prescription is in part derived from Alain Badiou's assertion in *Rhapsody for the Theatre* that "the pursuit of true justice is worth a bit of damage on the side of those who are well off."[5] But it also derives, I argue, from the fact that surveillance is a dialectical phenomenon. Inasmuch as it always produces its own dark shadow where transparency is lost and abuse of power is cultivated and thrives, so too, I argue, is the inevitable result of surveillance the production of its dialectical antinomy and thus of a response in kind—however asymmetrical that response might be. In this respect, the final chapter argues that postdemocratic formations not only call for but also cultivate the kind of theatrical or performative practice that is focused on direct creative confrontation and disruption of postdemocratic surveillance regimes instead of on the production of representational theater about surveillance.

The chapter finds historical precedence for this kind militancy in the work of groups like the Critical Art Ensemble and the Electronic Disturbance Theater, but ultimately it takes us back to the streets—the same streets where momentarily the book will begin its introductory reflections with a consideration of how the proliferation of surveillance technologies necessitate a substantial rethinking of the prescriptions for theater that Brecht laid out in his classic essay "The Street Scene." It is fitting then, that the final chapter of book also takes us back to the streets and to the creative political activism of two European artists: the German activist-artist Christoph Faulhaber, and the Russian dissident activist-artist Pyotr Pavlensky—both of whom have positioned their work so as to antagonize the authorities and both of whom have calibrated a provocative and sustained confrontational engagement with authority into the functioning logic of their wildly unpredictable performances. In Faulhaber's case this has led to works that breach the security parameters and protocols around U.S. embassies and consulates, and in Pavlensky's case it has led, among other things, to acts of arson that have targeted portions of the headquarters of the Russian Security Service. There is an uncommon courage and audacity in such works. It is the kind of courage that, I suggest, may be a fundamental requirement for artistic practice amid the post-democratic formations of the surveillance society.

If the book ends on a note suggesting that the political realities of the surveillance society necessitate a fundamental shift in our cultural orientation from the privileging of political art to a more direct commitment to a confrontational and disruptive creative activism, we might draw some reassurance from the fact that critics, practitioners, and activists have found themselves at similar junctures before. Walter Benjamin, for example, found himself at such a juncture midway through his most famous essay, "The Work of Art in the Age of Mechanical Reproduction," which was first published in 1938. Amid all of the talk of the loss of "aura," it is easy to forget just how radical his embrace of political art at the end of that essay actually was in its time.

Perhaps, for theater historians, this forgetfulness results because Benjamin's implicit point of reference in that embrace was Bertolt Brecht's theater, which is now familiar territory, and which for well over half of a century has proven to be such a rich resource of politically informed artistic practice that its sharp radical edge has been ground into safe domestic theater from overuse—not to mention the fact that, like the subsequent street theater of the Surveillance Camera Players, Brecht's edginess has also been eclipsed by the advent of advanced surveillance technologies. Keeping this latter point in mind, however, I want to suggest that one segue into the performative realities of the surveillance society leads to the streets where Brecht also headed to find inspiration for his notion of epic theater. But heading to those streets today, one will end up in a very different place than Brecht or Benjamin did as they walked the early twentieth-century streets of Berlin. Contrasting the realities of those streets with those of today is in fact a good introduction to how surveillance performs. So, all eyes to the streets!

From "The Street Scene" to the Seen Street

Among the many essays that Bertolt Brecht wrote on the politics and aesthetics of theater, his 1938 essay "The Street Scene," which was first published in 1950, stands out as among the most forward looking, even if that forward vision arguably overlooked as much as it perceived. The essay is perhaps best known for providing what Brecht in the subtitle called "a basic model for an epic theatre," and it derived its title from Brecht's primary example: a street scene from everyday life where a witness to a car accident uses a variety of gestures and provisional reenactments in order to demonstrate to bystanders the social significance of what transpired in the accident. The key moment in this famous essay occurs when Brecht lifts the performative logic of this scene from the streets and grafts it onto theatrical practice. Some streets lead to theaters, while others apparently reroute theater altogether.

Some sixty-five years later, Brecht's essay still has a lot of its cachet intact. At the very least, the essay offers a touchstone of late modernist aesthetics that continue to captivate the imagination of theater practitioners even today. But it is worth pausing long enough to consider what Brecht did and did not see in that street scene. Like so many victims in car accidents, there is a lot that Brecht never saw coming. In the broader scope of things to come, it may be that Brecht's eye on the street was itself more significant than his desire to bring the street into the theater. Indeed, within only twenty years of the publication of Brecht's essay, antiwar activists were admonishing the public not to consider the street scene as a model for theater (epic or otherwise) but rather to recognize, as the Living Theatre did at the end of *Paradise Now*, that "the theater is in the streets." What we discover in this closing slogan from *Paradise Now* is a subtle reminder that for all its conceptual rerouting of theatrical practice, the vision that emerged in Brecht's "Street Scene" essay directed readers away from the street and right back into that most conventional space of the performing arts: the physical building of the theater.

Of course, that building is a magical place. It is a place where scenes from the streets are immune from the passage of time. It is a place, for example, where one is as likely to discover Oedipus and Laius at the intersection where three roads meet, as one is to discover the Jets and the Sharks rumbling on the streets of New York's Upper

West Side. All this is possible even if actors follow the methods of epic theater and resist the temptation to draw their audience into the illusion of a realistic reenactment. But actual streets possess no such temporal immunity. Like all forms of infrastructure, they wear with time and need maintenance. They conform to the changing demands of traffic and commerce, and they evolve with the advance of technology. It is on such streets that Brecht's essay begins to show its age and, more importantly, that the street scene he envisioned in 1938 begins to show its surprising historical fixity. From a technological perspective, perhaps little had changed between the streets in 1938 and those of 1950 when Brecht finally published his essay—even amid the vast rebuilding of infrastructure that was necessary after World War II. But twenty-first-century streets are nothing like those that Brecht imagined in his essay and that serve as the backdrop to the model of epic theater that he proposed there.

Granted, however historically situated they may be, those imagined streets are still capable of illustrating the modernist precepts of epic theater. I am not contesting that. But ignoring the ways in which those same streets have become conceptual anachronisms in the present day avoids what is arguably the larger question raised by Brecht's eye to the street: the question about what models of theater and performance might emerge from a more dynamic understanding of streets moving and evolving with the flow of history. Such an understanding would position "the street scene" at an intersection where the conceptual spaces of twentieth-century political theater have long since yielded the right of way to the twenty-first-century technologies that are now integrated into the structural foundations of the public sphere as such. Simply put, this is an intersection where the "street scene" has given way to the "seen street."

If cultural theorists and critics, theater historians, performance studies scholars, practitioners, and theatergoers themselves are interested in serious rather than nostalgic debate about the contemporary viability of political theater in the public sphere—where the real battles of resistance and containment are being waged—then they cannot afford to turn a blind eye to the "seen street." Let them not delude themselves: today there are no intersections of consequence, no political crossroads or borders of significance, where an individual like Brecht's "street demonstrator" can improvise the particulars of an accident for curious bystanders without also doing so for the unwanted audience, that watchful eye of the advanced surveillance technologies which now routinely monitor the city streets.

In this new scene on the streets where actions fall under technology's watchful eye, there is something reminiscent of Walter Benjamin's discussion of Brecht's theories in his 1939 essay "What Is Epic Theatre?" There, Benjamin turns not to a street scene but rather to what he describes as a "family scene" that is interrupted by the sudden arrival of an uninvited stranger who is thus privy to the particulars of an altercation within the family's home. Benjamin never identifies this stranger or the authority that gives him liberty to breach the family's privacy, choosing instead to leave this dubious figure as an odd illustration of how epic theater favors "the representation of conditions" over "the development of actions."[6] Equally noteworthy in Benjamin's narrative is the absence of any clear indication of whether the family is aware of the stranger's presence. They are literally left in a state of suspended animation: something akin to a tableau vivant that the stranger casually surveils. No doubt, there is something theatrical in all of this, and at one level the stranger is an obvious personification of the audience breaking through the fourth wall, catching the family off guard and thus being in a position to take stock of their situation. Whether the stranger intervenes or simply observes and monitors is never clear. But that is secondary to the larger ambiguity of Benjamin's stranger.

This stranger might be the harbinger of a new theater—an epic theater—as Benjamin implies but never states outright. Still, the lack of clarity about him leaves the stranger as a potential harbinger of much more. With the advantage of hindsight, for example, the historian might catch a glimpse of an alarming aura of things to come hovering about this uninvited guest. Indeed, it is difficult to understand how, in 1939 with the Gestapo combing the streets and invading Jewish family homes, Benjamin could leave unaddressed the dubiousness of this figure or, by analogy, the implied complicity of the audience in his violation of the family's privacy. One needs to be careful, of course, in suggesting analogies between brownshirts and bourgeois audiences or in suggesting similarities today with the image Benjamin presented then as Germany and Europe were slipping beneath the dark shadow of fascist authoritarianism. A lot has changed since 1939—even though today one might legitimately see an unexpected foreshadowing in Benjamin's stranger of the immigration offices that are now disrupting the family homes of undocumented members of our communities.[7]

On that note, let me observe one significant difference between then and now: the uninvited intrusion no longer requires a stranger appearing at the door. Personification has given way to a proliferation of smart technologies: smartphones, smart homes, and smart streets render the stranger, anonymity, and privacy itself obsolete. But following Benjamin's logic, I want to suggest that even at the most basic level there is in fact a specter of fascism—or at least of autocratic and despotic control—in all of this smart technology as well. While many might argue, for example, that devices like smartphones represent an emerging counterbalance to the proliferation of video surveillance technologies on the street because of their own video function combined with direct access to social media, those phones possess a lot less progressive political potential than one might imagine. Here, video activists would do well to remember Benjamin's argument that "fascism sees its salvation in giving . . . [the] masses . . . the chance to express themselves" but only in such a manner that preserves existing "property relations."[8] Such relations survive even when more blatant acts of violence are captured on smart phone video—like the 2015 shooting of Walter Scott, the unarmed fifty-year-old African American who was gunned down in South Carolina by police officer Michael Slager after being pulled over for a nonfunctioning brake light on his automobile. We know about Scott because his killing was filmed by Feidin Santana, a Haitian immigrant, who, at considerable danger to himself, pulled out his cell phone, filmed the shooting of Scott by Slager, and then made the film public.

In that moment of spontaneity, Santana became part of something much larger than himself and larger than his individual act. Elise Morrison refers to acts like Santana's in her recent book *Discipline and Desire: Surveillance Technologies in Performance.* There in her own discussion of surveillance and the streets, Morrison notes that "there have been notable social movements [like Black Lives Matter] in recent years" that "use consumer surveillance, mostly through image capture in personal cell phones, to record and report police misconduct." Referring to "the widely publicized deaths of Trayvon Martin, Eric Holder, Michael Brown, Freddie Gray and a tragically growing list of other black men at the hands of police officers," Morrison notes that "visual evidence of these many deaths has often been provided through private cell phone video footage of fellow citizens"[9]—or, as in the case of Feidin Santana, of responsible and courageous neighbors who happen to be immigrants and who hope one day to become citizens.

But what were the consequences of Santana's courageous act "of conscience" that the *Washington Post* called nothing short of heroic?[10] The police officer, Michael Slager, was subsequently fired and charged with murder and, at the time of this writing, is awaiting a second trial, after the first trial against him resulted in a hung jury. But even if Slager is ultimately convicted in the second trial, the social order that produced him remains intact. Santana himself was quite aware of that social order even as he deliberated what to do with the video he had captured on his phone. As Jonathan Capehart reported in the *Washington Post*, Santana's act of heroism was in large part the product of a deep and understandable fear for his own safety after having filmed the shooting of Scott by Slager. In interviews on national television, Santana recounted that "fear for his safety almost led him to delete the incriminating recording and flee town." He explained that after trying to give the video "to the North Charleston police in [an] attempt to correct their false narrative of what happened," his fear for his personal safety increased and that ultimately it was a "fear of what the police might do" now that they knew he had the video that "moved him to get the video out"[11] to the public. Apparently, heroism has surprising sources and is often found in unexpected places. Santana, on the other hand, was easy to locate and still is. As a known person of color who challenged the police in South Carolina, his path now will never be an easy one, and the phone that he used to film the shooting of Walter Scott will help the police track Santana whenever they want.

There is no small irony that the very device that recorded the shooting of Walter Scott also increased Santana's chances of being subjected to police retribution. Indeed, amid all the celebration of Santana's streetwise heroism, it is easy to overlook how much more likely smartphones are to facilitate social control than to cultivate heroic improvisational activism. Whatever its video capabilities, the GPS technology on smartphones like the one used by Santana allows police to pinpoint individuals within fifty to a hundred feet of their actual location.[12] As Kai Biermann noted in 2011 in *Zeit Magazin*, smartphones, because of the way that they constantly access data, also create a record of an individual's movements and thus allow police to document an individual's whereabouts over a period of time.[13] All of this is possible with software technology that is increasingly available in the private sector. In this respect, the advantages of smartphones come at great cost, and I would suggest that smartphones like the one that Santana used arguably do more to maintain existing power structures than to challenge them.

The same can be said about what has been heralded as the one positive outcome of the senseless and unnecessary killing of Walter Scott in April 2015. His death has led to wider use of body cameras on police officers, which in practical terms means even more cameras on city streets. Elise Morrison has argued that while "ostensibly the increased visibility" resulting from body cameras ought to "improve police accountability and protect citizens," the use of body cameras "seems thus far to prove that evidence and justice function differently based upon the demographic and professional status of the eye of the beholder."[14] This is in part why the first trial of Michael Slager, for example, resulted in a hung jury. But we might push this notion of "the eye of the beholder" a bit further. Far from being the win that the news media celebrated it to be, there is reason to believe that the widespread introduction of body cameras was in fact a sleight of hand.

The indisputable result achieved by the introduction of body cameras was to move surveillance cameras beyond fixed locations and to make surveillance on city streets even more pervasive and intrusive. Many describe these body cameras as significant tools for achieving police accountability. The more accurate description of these tools would be something along the lines of advanced mobile CCTV cameras. After all, the cameras are pointed not at the police officers but at those who come within their visual horizon, and the implications here are alarming for life in the public sphere and on the city streets. With the ever-increasing sophistication of tools like face recognition software, it no longer matters whether one is a suspect or not. The cameras make a permanent record of who comes in the general vicinity of areas deemed in need of policing. Police who are deployed at political demonstrations, for example, are thus able to catalog at a whole new level of sophistication the identities of political activists, the frequency of their activism, and ultimately their prominence within activist communities.

But what about epic theater, and, to be more precise, what about the model of theater and performance that Brecht distills from that famous street scene—what happens to them in the onslaught of twenty-first-century technology and in the transition from the average street to today's smart street—the kind of street that is governed by "the culture of pervasive computing" and "digital technologies" that Jason Farman has argued are so deeply "woven into the fabric of daily life . . . that we are often rarely aware of the extent" of their "integration"?[15] Here flipping the phrase "street scene" and drawing attention to the "seen street" is about a lot more than a clever turn of phrase. It is foremost about focusing critical attention on how the street itself performs and on how that performance dictates the structures of just about any theater derived from the streets or about any theater in the streets that we might care to imagine. In this context, it is worth recalling Marshall McLuhan's general assertion back in the mid-1960s that "a new medium is never an addition to an old one, nor does it leave the old one in peace. It never ceases to oppress the older media until it finds new shapes and positions for them."[16] What I want to suggest through this reference to McLuhan is that the emergence of twenty-first-century technologies on city streets is more than merely "an addition" to them. Those technologies reconfigure the dynamics of the streets themselves.

Transposed into the particulars of Brecht's street scene, McLuhan's argument suggests, among other things, a reshaping and repositioning of what Brecht in his 1950 publication maintained was the "essential element" of the street scene: its "socially practical significance."[17] To be a model of theater, the demonstration that makes up the street scene must intervene "socially." This, Brecht maintains, is precisely what the street demonstrator does for the passers-by. His is not merely a provisional enactment. It is a demonstration with intent, purpose, and agenda. His demonstration is an intervention. Yet that essential element—the social intervention—is precisely what "the seen street" has usurped. Whether we like it or not, any subsequent intervention, any intervention along the lines described by Brecht, is now shaped by, indeed is always already positioned within, the frame of surveillance technology's sociopolitical interventions. But at some level, we already know this.

The entire CCTV industry is premised upon the recognition that surveillance technology is a form of social intervention, or of what the industry routinely labels "solutions." Chief among these so-called solutions is the marketed argument that the presence of surveillance cameras serves as a crime deterrent. A 2008 white paper from the company Motorola, for example, specifically tells potential customers that "the mere presence of video cameras is proving to be a significant deterrent to crime," and it claims that "neighborhoods where video is deployed have seen reductions in crime of up to 40 percent."[18] While there is ample evidence that such claims are exaggerated and misleading, there is little doubt about the significance of video surveillance systems as a social intervention that, to echo McLuhan, gives new shape and structure to human behavior and that, by extension, gives new shape and structure to the street scene as well. So while Motorola's claim that wireless CCTV systems are "transforming public safety" is debatable, the argument that such systems are transforming the public sphere is not.

One useful way to consider the broader implications of this transformation is to begin not with the actual CCTV cameras themselves but rather with their doubles, the dummy cameras that sell at a fraction of the cost of the real ones and that in many instances are indistinguishable from their expensive counterparts. I have to admit my own complete and utter fascination with these counterfeit surveillance cameras. In fact, I own a number of them. My fascination centers on their status as inanimate objects that simultaneously do and do not perform. What they cannot perform are the technical functions of an actual closed-circuit system of surveillance. But what they do perform are the functions of a portable system of theater. Wherever they are placed, these fake cameras masquerade as something they only resemble. More wonderful still, they arguably also perform their own small version of epic theater. At the most basic level, these counterfeits favor—and here I echo Walter Benjamin—"the representation of conditions" over "the development of actions." Most important of all, that "representation of conditions" is designed as a social intervention. This is, in fact, the primary value of fake CCTV cameras in the marketplace. On the packaging for one of the fake cameras I purchased, for example, the Chinese manufacturers claim that the presence of this fake camera will "help prevent or deter suspicious . . . activity." So too, according to the manufacturers, will it provide real "fun" to the owner as he or she watches "how friends and family react when they think they are being vedio [*sic!*] taped."

Immediacy 2.0 and the Rise of the Smart Street

This latter use of fake CCTV cameras may not conform to everyone's definition of fun, but it does serve as a reminder that surveillance cameras, whether real or fake, are never neutral objects in public spaces. They perform, and they generate new performances. Surveillance cameras affect not only how people act and perform in public spaces, but also how those public spaces themselves perform. Cameras affect their very substance and immediacy. Indeed, if there is an immediacy to those spaces—an immediacy to the streets that were once the setting for what Brecht called a "completely simple, 'natural' epic theatre"[19]—then the surveillance cameras populating city streets today, or to be more precise, the information technologies that comprise the very substance of those streets, deliver what we might call "immediacy 2.0." This is an immediacy where smart technologies from advanced CCTV cameras, biometric sensors, sound-level monitors, motion detectors, and cell phones have become, as Farman has noted, "absolutely integral to the very foundations of embodied space."[20]

Farman is certainly not alone in such arguments. He readily admits that his own work on mobile computing is indebted to the notion of "pervasive computing" explored by Howard Rheingold in *Smart Mobs* and also to the earlier notion of "ubiquitous computing" theorized by Mark Weiser in works like "The World Is Not a Desktop."[21] In many respects, these same debts find an additional echo in Malcolm McCullough's more recent book *Ambient Commons*, a work that, like Farman's *Mobile Interface Theory*, devotes considerable attention to mobile technologies as well. But the decisive juncture in McCullough's book comes with the argument that while mobile technology may remake "attention on the street," it is also on the street that mobile technologies meet "the embedded" and give way the kind of "urban informatics" that herald "a departure from [seemingly] all-powerful handheld gadgets." McCullough's book thus marks a significant shift from mobile interface into "embedded intelligence"[22] and into the kind of smart technologies that have not only become the building materials of city streets, commercial buildings, and family homes, but that have meshed them into complex grids of data flow and lucrative databases. To speak of "embedded intelligence" or of "smart technologies" is to speak of the increasing ubiquity of technologies that are designed to perform and, through their performance, designed to structure how citizens perform as well.

I will return to the notions of "pervasive" and "ubiquitous computing," "urban informatics," "embedded intelligence," and "smart technologies." But in order to explore the ways in which the seeming rhetorical neutrality of these concepts deflects attention from the harsh economic prerogatives that they reinforce, it is first necessary to garner some sense of the pervasiveness and momentum that these technological forces have acquired. Some sense of their ubiquity is conveyed in *Ambient Commons*, where McCullough likens the pervasiveness of smart technologies to "ancient Greek and Roman cosmologies in which mortals did not merely occupy their environment, but . . . were [literally] embraced by it,"[23] and inasmuch as that embrace today has supplanted cosmologies with powerful information systems designed to cultivate and harvest data, and thereby to shape human interaction and performance in the public sphere, there is something distinctly theatrical in the equation. Whether the point of reference is smart homes, smart buildings, or smart streets, McCullough argues that "a dramatist might understand this [embrace] as 'mise-en-scène.'"[24] That said, there is no exit—stage left or right—to this mise-en-scène. Citizens must follow the streets available to them—streets now layered in smart technologies that, according the McCullough, offer "no chance to opt out or even to tweak the resulting conditions. Your mere presence," he argues, implies "consent."[25] This is immediacy 2.0.

It is a restructured immediacy: an update on the old-fashioned immediacy described in Philip Auslander's 1999 book *Liveness*. Noting there that "media" and "mediation" are etymologically embedded in the term "immediate," Auslander argues: "Live performance is always already inscribed with traces of the possibility of technical mediation (i.e., mediaization) that defines it as live."[26] While Auslander's arguments occur within the context of challenging hard-and-fast distinctions between the performances of theater and the performances of television, the ubiquity of surveillance cameras and smart technologies in city streets signal a new phase in the very concept of immediacy as such. While there is still space in Auslander's argument for streets that accommodate Brecht's notion of a "completely simple, 'natural' epic theatre," the streets today lack that simplicity. They are always already inscribed not merely with the traces or with the possibility of technical mediation but with its constituting realities. Even here it is possible to dig a bit deeper into the realities of "immediacy 2.0," and to move from "the street scene" through "the seen street" into the substratic, performative realms of "smart streets" and the postdemocratic structures of what is called algorithmic regulation.

Performing Algorithmic Regulation: Ubiquitous Computing and Urban Fortification

In the final analysis, it is not so much the presence of cameras that constitutes the contemporary reality of city streets. It is the intervention behind the scenes of the cameras that constitutes that reality—the intervention of which the cameras are only a visible extension and symbol. In many respects, the cameras themselves are subsumed by more sophisticated and systematic technologies: the complex algorithms that process and correlate vast amounts of data being collected on city streets today by traffic sensors, GPS systems, smartphones, smart cars, and, of course, digital surveillance cameras.

The emerging norm of surveillance cameras today is an integration of them into the kinds of advanced software systems that now routinely provide "the detection and interpretation" services that were once the domain of skilled analysts in centralized control rooms.[27] Those analysts have been replaced by powerful computing systems processing big data from a range of sources that expand well beyond the visual field of even the most advanced CCTV cameras. With each passing day, with more and more data at their disposal and with increasingly sophisticated algorithms to process that data, these systems provide the kind of analyses of which their earlier human counterparts could only dream. Evegeny Morozov, for example, calls attention to the New York Police Department's recent implementation of what is called the "Domain Awareness System—which syncs the city's 3,000 closed-circuit camera feeds with arrest records, 911 calls, license plate recognition technology and radiation detectors."[28] Much of the focus of this system's algorithms is on providing real-time analysis based upon vast amounts of processed data. But other systems like the software called PredPol process even larger amounts of data and employ algorithms designed to predict areas where crime is likely to occur so as to allow the police to intervene with preventative actions.[29]

The predictive capabilities of software like PredPol highlight the increasing extent to which the algorithms employed today are capable of refining their own performances based upon new data (i.e., they are capable of learning). At the most basic level, they are also designed to intervene in public spaces, adjusting traffic light sequences, for example, when data indicates congestion, or reallocating police resources when areas are deemed "sensitive." On the face of things, examples like traffic regulation and preventative policing have genuine public relations appeal, particularly since they are based upon the collection of actual data. But as with the more tangible CCTV cameras, there is nothing passive or neutral about the algorithmic regulation that is central to these systems, nor about the data that these systems collect and process. On the contrary, algorithmic regulation is a high-tech form of automated social intervention on an epic scale. The question is whose interests these interventions reinforce, and here the signatures are fairly clear. Indeed, throughout this book I will suggest that these same signatures are to be found in virtually every form of surveillance in Western societies.

In the larger sociopolitical context, the emphasis on data collection and processing as well as the belief that solutions emerge when everything becomes quantifiable both belong to the basic principles of neoliberalism and to its underlying assumption that the best government is a government that not only reinforces the prerogatives of the business sector but that structures itself according to business standards of measured efficiency and performance. One need not look far to find advocates of this trend. In her recent paper "Why Data Must Drive Decisions in Government," Beth Blauer, who is the former director of Maryland's StateStat, admonishes state and local governments, for example, to "unlock and share their treasure troves of data"[30] with one another so that they will be able to catch up with the private sector, whose models of data collection and processing foster a model of efficiency that she believes the state should emulate, support, and serve. In such arguments, Blauer is but one voice among a chorus of those advocating a governmental shift toward open data and algorithmic regulation as a viable, indeed necessary, alternative to the inefficiency of local, state, and federal regulatory processes.

Among the most prominent voices in that chorus is that of the influential publisher and venture capitalist Tim O'Reilly, who not only champions "algorithmic regulation" as "an idea whose time has come,"[31] but who also draws his model of government directly from the business sector. "Just as companies like Google, Microsoft, Apple, and Amazon build [algorithmic] regulatory mechanisms to manage their platforms," he argues, "government exists as a platform to ensure the success of our society, and that platform needs to be well regulated" with algorithms that take the messiness of politics out of the equation.[32] But while O'Reilly and his cohort may see what one might call the "googlification" of everything as a panacea for inefficient government, cultural critics like Morozov are less sanguine about the automation of public policy and social intervention, particularly since the larger question of whose interests are being served[33] by such trends does not seem to be part of the basic script of algorithms. The "perennial problem of algorithms," Morozov argues, is "their presumed objectivity and quite real lack of transparency," and he concludes: "If no one can examine the algorithms—which is likely to be the case as predictive-policing software will be built by private companies—we won't know what biases and discriminatory practices are built into them."[34] Be that as it may, the biases buried and protected in the source codes of proprietary software have visible effects on the streets. There the binary codes of algorithmic regulation translate the numbers 1 and 0 into a clear division between the 1percent and the 99 percent, or more simply put, between "the haves" and "the have-nots."

Such divisions are a stark reminder that beneath the surface of seemingly innocuous terms like "ubiquitous computing," "urban informatics" and their euphemistic cognates are the technological foundations of what Mark Andrejevic calls "ubiquitous surveillance." Echoing David Lyon's notion that emergence of ubiquitous surveillance parallels that of ubiquitous computing, Andrejevic argues: "To the extent that networked forms of interactive communication become pervasive, they enhance and extend the monitoring process."[35] The question is to what end such monitoring occurs. Andrejevic rightly suggests that it has "speculative value for economic or policing purposes."[36] Keeping that dual value in mind, we begin to get a sense of the darker undercurrents of algorithmic regulation and the neoliberal policies that champion it. When we lose sight of "clear-cut boundaries between state and commercial surveillance," and between "policing and marketing,"[37] so too—ironically—do we begin to lose the basic structures of egalitarian democratic oversight. That which is considered to be antisocial becomes indistinguishable from that which is considered to be anticonsumer.

Across the city landscape such trends produce and regulate great economic divides. Critics like Pete Fussey and Jon Coaffee, for example, suggest that the path from ubiquitous computing to ubiquitous surveillance is really about "urban fortification." Unlike O'Reilly or Blauer, Fussey and Coaffee see the "embedding of surveillance cameras" and other smart technologies into "the urban landscape" as part of a multipronged strategy of reinforcing "structural inequalities" and of "'manag[ing] out' the urban poor or [other] 'undesirables.'"[38] This is especially evident in "urban renewal projects"[39] where the creation of smart streets is a key component in an economic strategy of contestation, displacement, and fortification. In this regard, the integration of city streets and neighborhoods into the surveillance grid and into the realms of what we might call "smart mapping" echoes a familiar historical narrative. The only thing new in this strategy is the smart technology that facilitates it. In *Ambient Commons*, for example, McCullough reminds us that "throughout its long history, mapping has seemed inseparable from colonialism," and colonial conquest. "To map a territory," he notes, "was to domesticate it,"[40] and here domestication is little more than a euphemism for appropriation and displacement.

On a smaller scale, "urban fortification" is a twenty-first-century variation of this same process, and it is a far more honest descriptor than, for example, the deeply problematic term "gentrification." The embedding of surveillance technologies into the urban landscape, as Fussey and Coaffee note, is typically part of a strategy to entice new businesses and investors into "formerly 'dangerous' parts of the city"[41]—"dangerous" being an obvious code word for minority and immigrant communities. Indeed, that enticement is consistently premised upon the also coded "use of surveillance cameras to gaze selectively on the conduct of specific groups of people" and thus "to keep city centers" and other sensitive areas "free of visible, less affluent and/or 'non-desirable' users."[42] Regardless of whether such measures lead to gated communities or to the so-called Ring of Steel[43] that cordons off London's financial district, there is a clear path from the Street Scene, through the seen and smart streets, to the Ring of Steel. The problem is navigating through the processes of social sorting that regulate access from one street to the next.

If Fussey and Coaffee voice concern over the larger urban spaces of surveillance, artist and critic Jordon Crandall brings similar concerns back to the individual streets. I draw attention to Crandall's work because, in essays like "An Actor of the Street: Events, Agencies and Gatherings," he focuses specifically on the evolving effect that algorithmic regulation has on the perception of normal and suspect activity in the city. Writing about the science of video analytics, Crandall speaks to the ways in which the algorithmic structures of advanced surveillance technologies establish and regulate social norms "through the categorization and standardization of information" and data—processes that are akin the social sorting that David Lyon has argued is typical of virtually all forms of surveillance. So too do these technologies define, identify, and regulate deviations from the very norms that their algorithms establish, assess, and constantly update. Indeed, when it comes to urban spaces and city streets, that regulation has direct relevance to the underlying assumptions of Brecht's "Street Scene," and here we come full circle in our discussion and arrive at a provisional conclusion that unfortunately is only diagnostic and offers no solution.

As Crandall notes: on contemporary city streets, video analytics assess what specialists call "dwell time" or "counterflow," and he explains that it is now commonplace for analytical programs running surveillance cameras to alert authorities to "people moving in the wrong direction," to people "loitering excessively," and to "the onset of abnormal densities of humans"[44]—all of which suggests that those involved in Brecht's street scene would be viewed with clear suspicion and, if not told to disperse, would be subject to further regulation and control. He explains furthermore that the purpose of assessing categories like dwell time and counterflow is to keep people moving and to keep them from amassing around sites and spaces of significance. Where Brecht's "completely simple, 'natural' epic theatre" goes amid such regulation is anyone's guess. Its fate is much like that of criminal activity in a surveillance society. Despite the claims of bureaucrats and businessmen, cameras don't reduce crime. They simply displace and move it to spaces deemed irrelevant or expendable. Along with that displacement,

CCTV cameras enhanced with video analytic software also squeeze the boundaries of acceptable behavior into an increasingly one-dimensional and uniform vision.

The practical realities of that squeeze have everything to do with the new regimes of ubiquitous computing and urban fortification. As streets become smarter and amass more and more data, the algorithms behind video analytics and behind a whole host of other smart technologies learn and adjust the parameters of their own objectives. They continue to refine in increasingly narrow terms the very categories of social norms that, ironically, their automated systems themselves establish. As a consequence, more and more forms of behavior on the street will fall under suspicion. The promise is that our streets will become increasingly fortified and sanitized, but so too will they evolve into spaces without incident or . . . scene. Programed for optimal performance, the street will be devoid of theater as we know it.

Ubiquitous Surveillance and Performance after the Epic Aura Fades

If this narrative of the transition from the street scene through the seen street to the smart street is a narrative of loss, perhaps there is an echo in the structure of that narrative that harkens back to previous narratives of loss and transition from which we might learn to look forward. And here once again, we encounter the long shadow of Walter Benjamin's highly influential meditation on the effects of mechanical reproduction on works of art. As I suggested earlier, Benjamin's meditation is of relevance to this project because he translates his initial narrative of loss into the historical foundation for the identification of a new and radical form of artistic expression that critics frequently have aligned with Brecht's notion of theater.[45] As is well known, Benjamin argues in "The Work of Art in the Age of Mechanical Reproduction" that the advent of mechanical reproduction resulted in the demise of uniqueness and authenticity as points of cultural and aesthetic value.

That value was significant enough that, in his essay, Benjamin elevates a work's uniqueness and authenticity to a quasi-mystical status. He likens a work's uniqueness and authenticity to an "aura."

A work of art's "aura," Benjamin argues, is "that which withers in the age of mechanical reproduction."[46] But it is not so much the loss of authenticity, that is, the loss of "aura," that is significant for Benjamin. It is that this loss of "aura" heralds a radical transformation in the status—indeed, in the very structure and substance—of art itself: "The instant the criterion of authenticity ceases to be applicable to artistic production, the total function of art is reversed. Instead of being based on ritual," Benjamin argues, "it begins to be based on another practice—politics."[47]

With this pronouncement, Benjamin's argument shifts from a lament on how the advent of mechanical reproduction has disrupted the authenticity of art into a meditation on how the advent of mechanical reproduction necessitates a retooling not only of artistic practice but also of the critic's understanding of what art is and what it does. In passing, it is worth noting the subtle way that this shift rehearses the Marxist adage that informs the first chapter of this book, namely that the beginning of all critiques is the critique of religion—a critique that I argue must also be extended to any serious critique of surveillance as a sociopolitical phenomenon. For if Benjamin's discussion of loss of "aura" does nothing else, it demystifies artistic practice and closes the chapter on the religion of art so as to make way for a critique that is unhampered by the unnecessary baggage of metaphysics.

If, for Benjamin, the loss of aura ultimately established the historical conditions for artistic practice like that which he believed to be exemplified in the work of Brecht, the advent of the digital age and of the surveillance society arguably sets us at a similar juncture with regard to our notions of political theater and performance more generally. Borrowing Benjamin's own language, then, I want to suggest not merely that Brecht's "street scene" has lost its aura or that the pristine image of a "completely simple, 'natural' epic theatre" is ill-suited as a model for effective theater and performance in the age of smart cities, but more importantly that history compels us now to look elsewhere for models of creative practice. Rather than lamenting this loss, I want instead to suggest that with the new realities of surveillance, smart technologies, and algorithmic regulation—all of which are embedded in their own conceptual models of performance—the base conditions of art-performance have fundamentally shifted. In simplest terms, that shift might be characterized as a transition from politically inspired art to artistically inspired political activism and civil disobedience: in other words, as a transition from art about conditions to the practice of artistic disruption in a social historical context where disruption must be state of the art in order to make a difference and effect change.

One might legitimately cite the work of the German artist Christoph Faulhaber in this larger call for creative disruption and artistically inspired political activism. He is one of the two artists whose work I discuss at great length in the final chapter of this book. As Ralph Eue noted in the catalog for the 2014 DOX film festival in Leipzig, where Faulhaber's film *Every Picture Is an Empty Picture* was shown, Faulhaber has very clear ideals about "what the 'venue' of art should be. His performances neither take shape in a safe studio nor are they presented exclusively behind the well cleaned windows of galleries. He prefers to stage his experiments against the order of the ruling systems right in the midst of society."[48] The film *Every Picture Is an Empty Picture* documents, among other things, Faulhaber's performances of *Mr. Security: To Serve and Observe* where in spaces under intense surveillance he staged his own acts of surveillance in order to challenge the legality of the prohibitions that ostensibly govern areas, for example, like that around the U.S. embassy in Berlin. Faulhaber's work attempts to recover public spaces from the logic of the surveillance industrial complex, and his work is informed by a sophisticated legal sensibility that allows him to challenge spaces that others understandably find intimidating. How effective his work proves to be is certainly debatable, but it is, I think, a step in the right direction, for it is cognizant of the scene on the streets today.

Moving further east and on to the streets of Saint Petersburg, one encounters similar work being done by the Russian dissident artist Pyotr Pavlensky, whose work I also discuss at length in the final chapter of this book. So in deference to those discussions, I will be brief here as I draw this introduction to a conclusion. In 2014, Pavlensky was arrested in Saint Petersburg in relation to a performance simply entitled *Freedom*. In many respects, that performance bore the hallmarks of the model of epic theater that Brecht proposed in his classic street scene essay, for it staged a social intervention that was itself a retelling of previous events on the streets. In this particular instance, however, the moment of representation was not alienated—at least not in the way that one normally associates with Brechtian aesthetics. Ostensibly, Pavlensky "and several colleagues reenacted pro-Ukraine protests by creating a barricade on Saint Petersburg's Tripartite Bridge, burning tires, and shouting."[49] But whatever echoes this act may have had of previous protests in the Ukraine, it was indistinguishable from what it also was: its own orchestrated act of disruption and of confrontation with the authorities, who moved in quickly, arrested Pavlensky, and thus set the stage for his chaotic yet intellectually sophisticated disruption of the Russian legal processes—a disruption that, as I will discuss later, literally flipped parts of the Russian security forces. Looking at works like *Freedom*, we quickly discern that Pavlensky's interventions tend to defy traditional categorization either as art or as protest. In this respect, they push the streets in new directions and have as their bearing what ultimately I will call "performance postdemocracy."

GOD'S EYES: CHAPTER 1

Of Sheep, Shepherds, and Oversight: Defrocking the Cultures of Surveillance

> Why has this church denounced and excommunicated theatre, thrown the actors overnight into common graves, and found luxury and oblivion in the public's zeal for the theatre spectacle, but not its own? Is it a matter of jealousy between tour operators, a sordid desire for monopoly?
>
> —Alain Badiou

A Pilgrim's Progress from Sociology to Surveillance: Reflections on Compartmentalizing Christianity

In the mid-1970s, InterVarsity Press published a short book entitled *Christians & Sociology: To the Challenge of Sociology . . . a Christian Response* (1975). At one level, the book, while now dated and out of print, is noteworthy as an example of the pseudoacademic books that InterVarsity Press published in the 1970s to support the evangelical campus ministries of the InterVarsity Christian Fellowship, which is a conservative Christian organization that focuses its ministries on university students. Indeed, reading through the book, one can easily see how it functioned as a valuable tool in those ministries. Time and again, the book explores the basic precepts of sociology (or at least its mid-twentieth-century formations) only then to circle back and subordinate those precepts to the assumed "revealed" truth of evangelical Christian doctrine. There is little question about the motivation behind the book's frequent return to biblical articles of faith. As the opening paragraph states, the book is designed to help Christians—and here the point of reference is clearly impressionable Christian students—who are young and "quite unprepared for the subtle and persistent tendency of sociology to erode faith and raise doubts."[1] Briefly stated, the kind of help that the book offers is to encourage its readers to place religious doctrine and academic discourse on equal footing and thus to position the metaphysical assumptions of evangelical Christian doctrine as one of sociology's legitimate options for explaining the world. The problem with this approach, however, is that in its efforts to provide Christian doctrine with a place among sociology's multiple critical narratives, the book consistently constructs a binary between Christianity and sociology as a whole—suggesting on the one hand, for instance, that "we could quite legitimately argue that sociology itself can be an ideology, [and thus] an example of false-consciousness,"[2] while claiming, on the other hand, that book's goal is to show Christians how they "can learn from . . . [sociology's] challenges . . . and how, in time, they might contribute usefully as sociologists."[3] But this formula is an equation of special pleading.

It is difficult to imagine how the book's readers would be able to contribute usefully as sociologists to a discipline that the book itself dismisses as an ideology and as "an example of false-consciousness"—unless of course the point of such dismissals is to clear the way for a reaffirmation of Christian beliefs as the foundational assumptions governing the practice of sociology or, for that matter, of any other academic discipline. This is in fact what book advocates. Not surprisingly, what thus begins in *Christians & Sociology* as a project to place religious doctrine and academic discourse on equal footing ultimately results in a subordination of sociology to the assumed primacy of revealed Christian truth. As it moves toward its closing arguments, the book tells its readers, for example, that "built into the Christian's sociology . . . will be the assumption that men in society are sinful, and that conflicts, deviance, even certain social institutions may exist as a result of sin."[4] In what is presented as a radical departure from the dominant secular assumptions of sociology as a discipline, the book invokes scriptural authority and finally concludes that "man is a slave to society, *not* because he is malleable, *or* because he is a social pawn, but because he is fundamentally a slave to sin, and therefore to a sinful society."[5] If one is familiar with fundamentalist Christian doctrines, there is nothing particularly startling in this conclusion. It is a variation of the same conclusions drawn by any number of evangelical writers, and it would not be of any real interest here were it not for the fact that the author of the book *Christians & Sociology* was a twenty-seven-year-old sociologist by the name of David Lyon, who is perhaps best known today (forty some years later) as one of the most prominent scholars in the field of surveillance studies. Indeed, Lyon is routinely cited along with Gary Marx, Clive Norris, and Gary Armstrong as one of the field's founding figures.

Drawing attention to Lyon's little-known, out-of-print book hardly outs him as a Christian sociologist. Evidence of his religious convictions is easy to find because it is as ample as it is open. Before turning his attention to issues of surveillance, for instance, Lyon published a second book with InterVarsity Press in 1979, this time a Christian response to Karl Marx, and he has since published two major studies of Christianity: *In the Steeple's Shadow* (Eerdmans, 1987) and *Jesus in Disneyland* (Polity, 2000). Whether writing on religion or surveillance, Lyon has consistently acknowledged that he "is guided by the converging traditions of Christian social thought"[6] and that his personal "convictions are Christian."[7] Almost the entire corpus of his scholarly work—which is prolific—contains references to those convictions, often, as Eric Stoddart notes, in "allusions or in footnotes to explicitly Christian texts."[8] I want to consider a number of those allusions at length later in this chapter, but I cite that first book from 1975 as my point of departure because of a particular concern that Lyon expresses in it about Christians who work within what he describes as "a compartmentalized mind" that separates academic inquiry from religious belief. Lyon states that such clear divisions do not strike him as "at all Christian" because "it means going back on the declaration of complete surrender that the Christian has made in response to the effectual call of the gospel."[9] But what, secular critics and scholars may ask, is the effect of such complete surrender? This is a question not merely about the answers one seeks in life or the conclusions one draws in scholarship (conclusion like Lyon's evangelical assertion that we are all slaves to sin). It is a question about how such "complete surrender" affects the very conceptualization of one's own field of inquiry as a scholar, and in this particular case, the question is not about sociology broadly defined but about surveillance studies as an emergent interdisciplinary field.

Given his strong Christian convictions, I want to speculate that it is not a matter of coincidence that Lyon was drawn to surveillance as a field of scholarly inquiry. For as a sociopolitical phenomenon—whether in the popular cultural imagination or in the practical hands of scientists, bureaucrats, and politicians—surveillance has consistently been conceptualized through the religious metaphors of the Judeo-Christian tradition. Sometimes these metaphors are subtler than others, but either way, long before Lyon began his work on surveillance, political philosophers, cultural critics, historians, and artists recognized the remarkable conceptual affinity between the idea of an all-knowing and all-seeing God and the idea of surveillance. Both are scripts of power and authority. To some extent, this means that the historical discourse of and about surveillance is marked in such a way that Lyon need not compartmentalize at all—he need not separate his academic inquiry from his religious beliefs—for to speak of surveillance is to speak within a discourse already embedded in the tropes and metaphors of his own faith.

Speculating about what attracts scholars like Lyon to the study of surveillance is one thing. Accepting his admonishment not to compartmentalize or separate an affirmation of the Christian gospels from scholarly inquiry is another matter altogether, particularly when surveillance is the subject of that inquiry—and this brings me to the focus of this chapter. Given the history of conceptualizing surveillance through the tropes and metaphors of Christianity, I want to suggest that Lyon's admonishment takes us in precisely the wrong direction. If the goal is a critical understanding not only of the techniques and technologies of surveillance but also of the ideologies that sustain them, then what is needed is a decisive separation of the critique of surveillance from a conscious or inadvertent affirmation of Judeo-Christian traditions. This is not to suggest that we should disregard those traditions. On the contrary, it is to advocate an open critical confrontation with them, the specific goal of which is a critique of the ways that Judeo-Christian traditions bestow ideological legitimacy and authority on the growth of surveillance in society. Without that confrontation and critique, studies of surveillance run the risk of reinscribing the very ideological structures that help to perpetuate and expand the surveillance society in the first place. The call here is thus as much for a critique of the ways that Judeo-Christian traditions have shaped key facets of surveillance studies as a field as it is for a critique of how those traditions bestow legitimacy on surveillance regimes in society more generally.

At a practical level, the question is how one might go about such a critique. As the coming arguments demonstrate, so much of the cultural history of surveillance as a concept is a history of tropes and metaphors appropriated from Christian discourse that it is difficult to disentangle the two. And yet my contention is that artists who have engaged those metaphors often display an amazing adeptness at extricating valuable critical perspectives on surveillance from that entanglement. In part, this is because their allegiances do not lie with the discursive economies of Christianity or surveillance, even though they often appropriate the former for their own creative and critical purposes. As the novelist William Gibson famously noted, "The street finds its own use for things,"[10] and, echoing that sensibility, one might extend it further. When the entangled discourses of religion and surveillance hit what we might call "the streets of culture"—the streets paved with the work of artists, wordsmiths, and performers—they are retooled and redirected into unorthodox yet extremely valuable critical visions. The kind of unorthodox visions I have in mind here are those that emerge amid the strategic ambiguities of artistic expression and that thus contrast with the orthodox notions of absolute truth and divine omniscience in Christian doctrine.

A Divine State of Affairs: Surveillance, Secular Drama, and Church as State

Often those moments of ambiguity play out against the backdrop of narratives that initially appear to offer powerful critiques of state-sponsored surveillance by characterizing it as an arrogant grasp at something akin to a godlike power. In many respects, this characterization works well because it taps into an established binary between the church and the state that appears to facilitate a critique of surveillance that is grounded in the moral authority of the church. But in works with no clear commitment to the underlying assumptions of Christian doctrine, the critique of surveillance frequently emerges from a creative deconstruction of that binary. Framing their visions of surveillance within the tropes and metaphors of Christianity, works like Muriel Spark's *The Abbess of Crewe* and Arthur Miller's *The Archbishop's Ceiling*, for example, offer subtle critiques of the institutions of Christianity that, in turn, serve as the foundation for a more penetrating critique of surveillance itself. Both works merit further consideration since the issues that they raise resonate with those that are examined throughout this chapter.

Spark's novel and Miller's play were both published against the backdrop of the Watergate scandal, and while they frame their considerations of surveillance within images of the church, its institutions, and its discourses, the critical rupture of the church-state binary results from the fact that in both works those images serve as metaphors for, rather than as counterpoints to, state corruption and abuse of authority. As a consequence, the church is never quite removed from the object of the critique that it facilitates. Blurring into images of the state, it is sullied, as it were, with guilt by association. This blurring is perhaps most obvious in Spark's novel *The Abbess of Crewe.*

The plot of Spark's novel is a relatively simply allegory that recasts the events of Watergate as a struggle for power within a convent that falls into scandal and crisis when it is discovered that its newly elected abbess, Sister Alexandra, resorted to bugging the convent so as to guarantee the defeat of her main rival for the position of abbess, a younger and more progressive nun by the name of Sister Felicity. In an obvious allusion to the Watergate break-in, the plot against Sister Felicity, while effectively ensuring her defeat in the election, comes to the light of day when it is discovered that one of Sister Alexandra confederates broke into Sister Felicity's private files. As was the case with its historical counterpart, the scandal erupts not so much from the break-in as from its attempted cover-up.

While its obvious parallels to Watergate certainly invite readers to view the wiretappings, buggings, and burglaries of the Nixon campaign with moral indignation, *The Abbess of Crewe* leaves little moral high ground for the church itself or for a notion of surveillance that is conceptualized through metaphors drawn from its doctrines. Inasmuch as Spark's novel positions the church as a stand-in for a government not only intent on using surveillance technologies for self-serving ends but also intent on appropriating established political discourse to mask its abuses, *The Abbess of Crewe* leaves its readers with two specific critical issues to consider. First of all, the novel asks readers to consider how the church and its doctrines might legitimize and provide cover for practices comparably dubious to those of the Nixon administration; and, as a result, so too does the novel challenge the reliability of the images and doctrines of the church to provide a viable critical understanding of—or a stable moral or ethical vantage point from which to view—surveillance as a sociopolitical phenomenon.

Nowhere is this latter challenge more evident than when the newly elected abbess, responding to an inquiry from the Sacred Congregation for Religious and Secular Institutes in Rome, writes a letter to its prefect, explaining in a nondenial denial of her own activities "that electronic surveillance (even if a convent were one day to practice it) does not differ from any other type of watchfulness, the which is a necessity of a Religious Community."[11] The rich ambiguity of this statement lies in the fact that it opens an aporia in the Christian doctrine in which it participates. Read carefully, it suggests not only that surveillance is a modern form of Christian watchfulness, but also that Christian watchfulness is no less dangerous—no less inherently susceptible to a use contrary to its own rhetoric—than surveillance itself. That susceptibility is an always-present part of the discourse. Indeed, the value of the abbess's letter over and above its immediate relevance to the novel's plot is that it points toward an understanding of Christian discourse beyond notions of use and abuse, of purported intent and its opposite, or even of good and evil. It suggests that both use and abuse are in fact figured into the same moment: that what, as a matter of social convention, we label as "abuse" or "misuse" of Christian discourse is always an intrinsic part of what the discourse facilitates and produces and always an inherent part of how the discourse itself performs. Inasmuch as the novel conceptualizes surveillance through the tropes and metaphors of that discourse, *The Abbess of Crewe* implies that the same is true of surveillance as well.

Although there is palpable satire in Spark's narrative account of the abbess's self-serving contextualization of surveillance within the doctrines of the church, one of the most striking aspects of her letter is how its clever deconstruction of Christian discourse has escaped otherwise insightful readings of *The Abbess of Crewe* by critics like

Lewis MacLeod. I cite MacLeod here because he specifically draws upon David Lyon's 2001 book *Surveillance Society* in his study of Spark's novel and because he uses Lyon's definition of surveillance as having "two faces"—one looking toward "care" and another looking toward "control"[12]—as the basis for positing a distinction between what he suggests are two competing forms of omniscience in the novel. Even though I question the clean delineations in MacLeod's argument that Spark associates God's omniscience with "care" while associating human efforts to gain omniscience with the quest for power and "control,"[13] his argument highlights a more important issue about Lyon's characterization of surveillance as having "two faces," namely that those faces are conceptualized as binary alternatives: as "good" surveillance versus "bad" surveillance, or, as the case may be, as "good" omniscience versus "bad" omniscience. Inasmuch as the operative assumption beneath those two faces is that with regard to surveillance we can and should move toward the good and leave the bad behind, both Lyon and MacLeod overlook what is arguably one of surveillance's defining characteristics: that it always unavoidably accommodates—indeed, simultaneously facilitates and produces—the bad in pursuit of and along with what is promoted as the good.

At a conceptual level, there are no clear alternatives or choices between good and bad with regard to surveillance. But this does not mean that surveillance is somehow neutral. It is rather to suggest that the bad is always figured in as a possibility, if not an actual probability, along with each instance of the good. Visualizing surveillance through the lens of Christian discourse obscures this fact because Christianity itself is arguably entangled in a comparable dynamic that its doctrines obscure. In the logic of Christian beliefs, the godly ideal is held as an always inviolate and desirable spiritual alternative to dubious secular ambitions and actions. Since the ideal in this belief system is never measured against its actual manifestations—and only the manifestations are measured against the ideal—the binary and its privileged signifier always retain their inviolate status regardless of what they actually produce. In light of this dynamic, the problems with MacLeod's arguments about "care" and "control" in Spark's novel tell us as much about conceptualizing surveillance through the lens of Christian discourse as they do about *The Abbess of Crewe*.

To associate surveillance with omniscience is to set its very definition in a measurement against inviolate Christian ideals. This is also true of conceptualizing surveillance in a binary between care and control. At an obvious level, this conceptualization suggests that there is a godly omniscience of care against which a secular omniscience of control can be measured and against which it always comes up short. But the problem is that this binary is not limited to a contrast between a Christian notion of omniscience on the one hand and a secular notion of surveillance on the other. The binary slips into the conceptual foundations of surveillance itself, providing it with its own variation of an inviolate ideal that justifies surveillance's continued development and use. Here the binary between care and control is no longer merely a contrast between divine omniscience and secular surveillance but is extended as the "two faces" of surveillance: a surveillance of care and surveillance of control. In this latter variation, the notion of a surveillance of care (or, as the case may be, a surveillance of security or of oversight) becomes the privileged signifier that retains its inviolate status regardless of the results that surveillance actually produces. And this brings us back to what, in contrast to MacLeod, I have argued is the major significance of *The Abbess of Crewe*. Rather than reinforcing the binary of care and control, Spark's novel ultimately exposes the extent to which modeling an understanding of surveillance after the Christian notions of "watchfulness" or "care" rationalizes away and thus provides cover for the potential for abuse and corruption that the powerful technologies of surveillance always produce despite rhetoric heralding surveillance, ironically, as a countermeasure . . . to abuse and corruption. We will return to the issue of "care and control" toward the end of this chapter.

As an allegory of the Watergate scandal, *The Abbess of Crewe* focuses on the use and abuse of surveillance by figures of authority. Arthur Miller's play *The Archbishop's Ceiling*, on the other hand, takes up the issue of surveillance's effects on "the surveilled," shifting critical focus from figures of authority to those who rightly or wrongly believe themselves to be the targets of surveillance. Briefly stated, Miller's play takes place in an Eastern European capital (presumably Prague) at the height of the Cold War. It centers on events in the former residence of the archbishop, which has a magnificent fresco ceiling that reportedly has been bugged by government officials, and which now has as its occupants two mildly successful writers by the name of Maya and Marcus. The two writers are also government informants/operatives, and the plot revolves around their subtle efforts to obtain compromising information on the country's most famous dissident author while receiving an unexpected visit from an American author.

Unlike Spark's novel, Miller's play is thus less concerned with actual instances of surveillance than with the complicated effects that its mere possibility has on human interaction, and in his drama, these concerns translate into what I would argue is a Cold War allegory of the life of dissident artists in an Eastern European police state—an allegory that in name is about Eastern Europe but that in spirit and implication also expresses deep foreboding in the aftermath of Watergate about the increasing drift in the West toward a surveillance society. As is the case with Spark's novel, the concept of surveillance in *The Archbishop's Ceiling* is drawn in a problematized Christian metaphor, but in this particular instance, it is reduced to a single overdetermined image. Miller distills surveillance into a powerfully ambiguous image of a bugging device that state police may or may not have hidden among the cherubs in a ceiling fresco at the former residence of an archbishop that is now "frequented by local and foreign writers."[14] As Andrew Sofer has noted, at some level "the bug [in Miller's play] has displaced God as omniscient, inscrutable observer and judge of human behavior"[15]—a point that the presumed existence of the bug in the painted heavens above would seem to make abundantly clear. But just as no one knows for certain whether God is actually taking note of our actions or if he even exists, none of the characters in *The Archbishop's Ceiling* know whether they are actually being surveilled with the bugging device allegedly hidden in the ceiling fresco.

Critics have understandably argued that the effect of this ambiguity within the plot of Miller's drama is something akin to that of Jeremy Bentham's notion of the panopticon, where, in theory at least, prisoners are never quite certain whether at any given moment they are being observed and thus, always fearing punitive reprisal for recalcitrance, conform their behavior in accordance with expectation. Similar to Michel Foucault's extension of this idea to society more generally, Miller himself has suggested that "we are all secretly talking to power, to the bugged ceilings of the mind."[16] The implicit logic of this assertion is that it doesn't matter whether the bug is actually hidden in the ceiling. The mere potential that it might be results in what Sofer, echoing Miller, describes as an image of "authoritarian power that has so saturated the self that to some extent we are all speaking [as Miller suggests] to the bugged ceiling of the mind" and thus also conforming our behavior to the assumed expectations of a hidden audience.[17] Inasmuch as such arguments position the self in an anxious relation to an unseen audience, they locate the image of state surveillance in Miller's play at the crossroads of theatrical and Christian metaphor—similar to where Bentham himself, not unproblematically, located it.[18] But so too do they conceptualize the image of surveillance in *The Archbishop's Ceiling* within a church-state binary that largely relies on a generic notion of the state (even if it is identified as a police state) that limits the complexity of the vision of surveillance that actually emerges from Miller's play.

I want to argue by contrast that the complexity of that vision is specifically linked to Miller's decision to set his drama against the backdrop of intellectual disillusionment with—indeed, against the backdrop of what we might cautiously call "a loss of faith in"—Soviet-style Marxism. That loss of faith, I argue furthermore, not only complicates a simple framing of the scholarly discussion of *The Archbishop's Ceiling* with a church-state binary. It thus also exposes the limitations of framing the image of surveillance in *The Archbishop's Ceiling* within Christian metaphor. But understanding the image of surveillance beyond those limitations necessitates a closer look at the dynamic of the church-state binary in Miller's play.

Sofer is certainly not alone in his reading of Miller through the lens of that binary, a reading that Miller's own reflections on the play admittedly don't discourage. Christopher Bigsby, for example, also embraces this approach in his book *Arthur Miller: A Critical Study*. His arguments are worth considering in some detail because they so clearly plot out the ways in which the antinomies of that binary are figured into the play's implied point of reference: the ceiling itself. Bigsby suggests that, with its heavenly images of cherubs, the allegedly bugged ceiling of the archbishop's former residence has a dual function. First of all, its baroque fresco is a reminder of an earlier time when "human life was charged with significance by the conviction" that we all "acted out our dramas in the eye of God," who then judged us in terms of "grace or damnation."[19] But with the advent of a new political order, not only did the archbishop vacate the residence; a new ideology also supplanted the metaphysical order that the archbishop represented.

Following this argument, the archbishop's ceiling can be understood as a site where those two orders blur since it contains simultaneous references to both. Thus, Bigsby identifies the ceiling's infamous bug as the symbol of the power structure that filled the void left when the archbishop vacated the residence. Fitting the image of state surveillance into the conceptual imprint of a previous Christian order—and hence modeling his interpretation of surveillance after that order's imprint—Bigsby argues that the ceiling also serves as a reminder of a void that surveillance presumably fills now that God is "no longer there." The blurring of these two orders, so Bigsby suggests, is evident in the fact that while one has given way to the other, we are all still playing to "a hidden audience."[20] In terms of its existential effects, the new audience is thereby structured, according to Bigsby, very much like the one of old. "To be bugged is to be violated," he argues, "but it is also to have one's existence confirmed and one's meaning attested."[21] Yet introducing the idea of existential confirmation in this context is, I would suggest, to place the explicit imprint of Christian metaphysics on the new order rather than to have the new order actually supplant the previous one.

While theater historians and performance studies scholars might delight in the idea that an audience—even a hidden audience—is so essential to existential confirmation that individuals are willing to suffer violation in order to obtain one, the operative dynamics of surveillance are lost in the notion that God's absence is somehow so traumatic in effect that it inclines everyone to suffer violation willingly in order to find substitute meaning through surveillance's technologies. Indeed, comparing a person's response to surveillance with a quest for existential confirmation in the absence of God overlooks the extent to which *The Archbishop's Ceiling* actually chafes against the notion that surveillance is a rough surrogate for Christianity's divine but hidden audience. As an alternative, I want to suggest that the basic church-state structure in Bigsby's argument not only positions the church as the privileged signifier in its binary and thus offers an image of the state that always already bears the church's conceptual imprint. It also overlooks a crucial third element in the mix: the philosophical backdrop to Miller's play. That backdrop comprises an uneasy tension between a defunct Christian metaphysics and a Soviet-styled Marxist ideology with which all of the play's central characters are disillusioned. I want to return to that disillusionment momentarily, but first there is the matter of its conceptual context.

The church-state binary in Bigsby's argument is in many respects a scaled-back version of the tense philosophical divide between religion and Marxism. Much of that tension results from their largely irreconcilable philosophical outlooks. On the one hand, there is the classic Marxist critique of religion as a form of alienated consciousness or, more simply put, as an illusory response to real suffering. The basic script of this critique is easily summarized: "Religion expresses the real distress and suffering of the people who live on earth yet seek their identity, consciousness and self in the beyond"[22] rather than attending to the historical material sources of their suffering. The conservative response to this critique has been to accuse Marx of denying "religious transcendence" while keeping "transcendence" as a utopian ideal.[23] In simpler terms, the charge is, as Bertrand Russell noted in the 1940s, that Marxism has "all of the characteristics" of a religion.[24] Adopting this same perspective in the book that he wrote on Marx for InterVarsity Press, David Lyon argues that "Marx's error was to imagine that he was not making a religion himself,"[25] an error that Lyon uses to reassert a "biblical perspective" on the critique of "industrial capitalism."[26] My sense has always been that this critique of Marx mistakenly equates utopian ideals with the assumption of divine agency, but inasmuch as this critique suggests that Marx's philosophy carries the imprint of the order that it presumably replaces, there is a striking similarity between the conservative critique of Marx and Bigsby's interpretation of the significance of the ceiling in Miller's play. For both conceptualize religion as a privileged signifier that is reproduced in the order that presumably replaces it.

Returning to Marx's critique of religion as an alternative to Bigsby's approach has the advantage of offering the opportunity to consider the ways in which the bugged ceiling in Miller's play suggests parallels between surveillance and religion rather than suggesting that surveillance is modeled after religion or is a surrogate for it. Following this line of thought, my contention is that the effects of surveillance are only like those of religion in that both offer unique examples of what Marx identifies as alienated consciousness. Like the effects of religion, the effects of surveillance provide an illusory balm for genuine human suffering. While religion promises solace and meaning in a world whose material realities produce disenfranchisement and alienation, surveillance promises security and stability in a world whose material realities breed the threat of violence, abuse, and oppression. But surveillance technologies are as much a part of those threats as they are a solution to them, and in this respect, the false consciousness encouraged by surveillance—its cultivated illusory complacency and its false sense of security—suggests the need to question the ways in which faith in surveillance distracts from the root causes of the very conditions to which it seemingly responds. As Bigsby rightly notes, "To be bugged is to be violated," and more than anything else, *The Archbishop's Ceiling* provides a historical marker of that sensibility. Its characters offer a glimpse into a moment when political material realities were so configured that there is little if any illusion about the purpose of surveillance or about the nature of the political forces behind it. In *The Archbishop's Ceiling* the surveillance society cultivates a culture of disillusionment.

The vision of surveillance's effects in Miller's play is inseparable from the dynamics of that disillusionment. Not only does it lead to numerous instances of subtle rebelliousness in the play that challenge the self-imposed conformity that is the supposed result of the hidden audiences of Bentham's panopticon and Miller's "bugged ceilings of the mind," but it also suggests that the notion of an audience does not translate as fluidly as Bigsby suggests from the idea of a watchful and judging God who knows everything to that of an eavesdropping surveillance technology that obtains raw information in need of interpretation. In this respect, the ceiling is a bit of a ruse. The action is not in the ceiling above but in the intrigue below. What transpires beneath the ceiling of the archbishop's residence in Miller's play offers a very different model of surveillance—one where that which transpires on the ground stands in marked contrast to that which is ascribed to it above in the Christian iconography and metaphor of the ceiling. The clearest sense of what that model entails emerges from the fact that in Miller's play, the alleged bug in the ceiling is not some sort of omniscient or omnipresent force, monitoring the day-to-day affairs of all within the state. Citizens have to be lured within its range, and it is only one aspect of a larger ploy. It is a technical device within a nefarious political drama staged within the archbishop's former residence—a drama that is rightly called "very bad theatre"[27] by the character Sigmund, a dissident writer who is probably modeled after Václav Havel and who is presented in the play as the country's greatest living dissident writer.

Many would argue that the plot of *The Archbishop's Ceiling* centers on Sigmund. But the play plausibly centers on his two contemporaries instead: the mildly successful Eastern European writers, Marcus and Maya, who lure other writers—particularly those critical of the government—to the archbishop's former residence, where they ply them with alcohol and sex in order to lead them into compromising behavior that, once recorded, can then be used against them in state-sponsored acts of manipulation and blackmail. What makes this stratagem such bad theater is that its justifications are little more than shoddy masquerade. While it is done in the name of state security, those who execute the ploy do so not out of sense of ideological conviction or political allegiance to the state, but rather for reasons that are self-serving, opportunistic, and entrepreneurial—like the privilege of living in the residence itself, having a passport, and being able to produce cultural programs on the radio.

All of this is made clear when a somewhat naive American writer by the name of Adrian arrives unexpectedly at the residence. The play opens with him in the setting room of the residence waiting for Maya to return with coffee and nosing around the room looking for the bug. Indeed, when Maya returns, he confronts her with the rumor that the "house is bugged" and accuses Marcus and Maya of "bring[ing] in girls to compromise writers with the government."[28] Although Adrian himself is never really a target of these staged entrapments, he is on very good terms with Sigmund, who, as it turns out, is Marcus and Maya's primary target and who, much to Adrian's surprise, shows up at the residence with Marcus and a Danish woman by the name of Irina. Once Sigmund arrives, he begins to realize that Marcus and Maya's goal is to convince him—presumably at the government's behest—to leave the country and live in exile. As Sigmund resists and it becomes increasingly clear that, despite substantial pressure, he is unwilling to go into exile, desperation ensues, and, in what is arguably the play's most significant contribution, a bleak image of surveillance as an ever-expansive active force of co-optation emerges from the plot.

That image has very little to do with the Christian iconography under which it emerges, and it has even less to do with Sigmund than with Marcus and Maya, who by the play's end arguably have become harbingers of a cynical pragmatism that the play posits as the effect of surveillance on social interaction. The dynamic here is easy to identify: the more Sigmund resists their efforts to force his exile, the more Marcus and Maya become anxious about what they themselves say in hearing range of the bugging device, which rather than surreptitiously invading their privacy is in fact functioning with their full knowledge and provisional consent as a trade-off for the privileges they enjoy. For all intents and purposes their privileges have come with a terms-of-service agreement. Rather than emulating the mannerisms of an omniscient God, the bug in the ceiling functions more like a corporate manager concerned with quality control and thus with measuring Marcus and Maya's performance with respect to how effective they are in delivering Sigmund. But with regard to issues of human interaction, the more general principle behind the bug's surveillance of Marcus and Maya, I would suggest, is an ongoing assessment of how well they have been integrated into, and how effectively they enhance, the ever-increasing expansion of the surveillance apparatus of the state, an expansion that is arguably as much of an end in itself as it is a means to any particular end like actual state security.

As is the case with virtually all forms of surveillance, that apparatus is an active rather than a passive force. It instigates rather than merely records, and orchestrates rather than merely observes. In this respect, Marcus and Maya's manipulative behavior toward the writers whom they host at the residence is merely a mid-twentieth-century foreshadowing of a surveillance dynamic that has evolved with increasing technological sophistication as we have moved through twentieth-century Soviet-style authoritarianism into twenty-first-century global capitalism. Indeed, it now is normalized and commonplace: perfected not behind the Iron Curtain but on the wide expansive stage of global capitalism. What we thus encounter in Marcus and Maya are not simply subjects of surveillance but also a dramatic illustration of how being constituted as subjects of surveillance effectively renders individuals into its active agents as well. Cynical as Marcus and Maya may seem in Miller's play, they are the forerunners of the future surveillance society, and at some basic level they are little more than unrefined prototypes of the groomed customer service representatives we now encounter on a daily basis with whom interaction is prefaced by notification that "this conversation may be recorded for purposes of quality control."

For his part, Sigmund offers what might be described as the fading critical voice of a disappearing era. Calling the American writer Adrian a "scientist observing the specimens," Sigmund gestures "toward the ceiling" as he tells him, "They are killing us, Adrian—they have destroyed my friends!"[29] In simplest terms, that destruction amounts to being swallowed into shadows of inescapable disingenuousness where, to borrow the colloquial negative inflection of the term, all is "performance." Sigmund hints at something like this earlier when he explains to Adrian the effects of a surveillance state on human interaction: "We . . . believe we can escaping [*sic*] power—by telling lies. . . . We must lie, it is our only freedom. . . . Our country is now a theatre, where no one is permitted to walk out, and everyone is obliged to applaud."[30] This is a theater where the audience, like the bug in the ceiling, is implied and unseen; where one survives with a wink and a nod while paying lip service to that which one does not believe; and, where beneath the wink and a nod, some semblance of truthful human interaction presumably survives. But I would suggest that through the example of Marcus and Maya, Miller implies that this survival is temporary at best and that in the shadows cast by the unseen audience—the shadows presupposed in Sigmund's explanation—that which is said slowly loses decipherability as the authentic product either of conviction or of adherence to scripted expectation. The wink becomes no longer clear, and the nod becomes not quite discernible, and Sigmund's implied notion of covert authenticity blurs into ambiguity and cynical equivocation.

There are many open questions at the end of *The Archbishop's Ceiling*, but the chief among them concerns Miller's uneasy framing of this narrative of surveillance in Christian metaphor. To some extent that framing has a lot to do with Sigmund's notion of covert authenticity, which is akin to the Christian notion of truth. But so too is this metaphor connected to his status as a dissident and with the etymology of the term "dissident" itself, which in a striking instance of serendipity links the character Sigmund—as a dissident—with all that is figured into the ceiling fresco at the archbishop's former residence: its current symbolic and probably literal function as the reminder of state surveillance and its earlier artistic function as the reminder of God's omniscience. In fact, the notion of a dissident as we understand it today first emerged in the 1940s as a term identified with those who opposed the Soviet-style authoritarianism that Miller distills into that bug in the ceiling. But prior to that, the term "dissident" was first used to identify those critical of the church and of the hierarchies that Miller summons with the image of ceiling fresco and the residence itself. At a conceptual level then, Sigmund is positioned—specifically as a dissident—in a dialectical relationship to the historical trajectory of the archbishop's ceiling in Miller's play. As a dissident, he is its historical counterpart, and as a consequence he and the notions of truth and authenticity he implicitly defends are also entangled with the Christian metaphor that initially suggests but that ultimately distracts from the final distressing image of surveillance that emerges from the play.

That final image is easily lost in the powerful diversion created by play's title and by its suggested equation of surveillance with the theological views associated with Christian iconography. Indeed, while virtually all of the characters repeatedly point toward the alleged bug in the ceiling, the play is implicitly pointing elsewhere: toward Marcus and Maya, who facilitate and expand the surveillance apparatus irrespective of whether the bug is present. If the overriding argument about the bug is that it doesn't really matter whether it exists, the reason it doesn't matter is not so much because of some notion of "bugs in the mind" but because Marcus and Maya are doing the bug's bidding with or without it. In this respect, it is worth considering what they represent in relation to the probable historical figure after whom Miller modeled Sigmund: Václav Havel.

In an insightful reading of this probability, Christopher Bigsby notes that "if Sigmund was indeed based on Vaclav Havel, as Havel himself believed, the rebel against the state became the state," and while "state repression ended," it was replaced, Bigsby rightly observes, by the "new imperatives" of "international capitalism and its products."[31] But, as we now know, those new imperatives did not do away with the bugs in the ceiling; they allowed private enterprises to replaced them with more advanced technology and to monitor citizens for profit. For among the most prominent products of international capitalism are the technologies and techniques of surveillance, which in the years following the Velvet Revolution that brought Havel to power have bourgeoned into one of the most profitable and promising industries in the global marketplace. There is no ceiling for Marcus and Maya in such a marketplace. The sky's the limit—whatever the cost to human interaction. In this respect, the effect of surveillance on human interaction is not redemption but reification.

Of Eyes, Art, and the Prerequisite of Every Critique

In the critical dialogue that they create with Christian discourse, works like *The Abbess of Crewe* and *The Archbishop's Ceiling* challenge the simple equation that surveillance is something akin to lust for godlike power, and they set a valuable precedent for a critique of surveillance more generally. There is an echo in that precedent and even inspiration to be drawn from the young Karl Marx, who in the opening passage of the 1843 introduction to his "Critique of Hegel's 'Philosophy of Right'" famously argued that "the critique of religion is the prerequisite of every critique."[32] Almost 175 years later, I would contend that if the general validity of Marx's assertion has diminished at all with the passage of time, it is nonetheless still valid for the critique of surveillance in particular. The critique of religion is requisite for the critique of surveillance because without it the temptation will always be to assume that the latter is some cheap counterfeit of the former. Given Lyon's profession of faith in the preface to his first book on surveillance, which is entitled *The Electronic Eye*, it ought to come as no surprise that the book succumbs to this very temptation. As Stoddart has noted, Lyon likens the development of the "multitudinous technologies" of surveillance to a quest for "secular omniscience," a quest that Lyon rejects as a "false faith."[33] This contrast between the ideas of an authentic faith and a false one is particularly evident in the book's implicit defense of a unified notion of the soul—a defense that echoes Lyon's earlier dismissal in *Christians & Sociology* of scholars who argue "that people do not have a 'self' but rather a succession of roles."[34] *The Electronic Eye* may be more nuanced in its argumentation than that early work from Lyon, but the underlying assumption about souls and their essential sinful nature is the same.

That assumption is *The Electronic Eye*'s point of departure. Its cover, for example, dons a reproduction of René Magritte's 1929 painting *Le Faux Miroir* (*The False Mirror*), which like so many of Magritte's works, is a kind of elaborate pun that in this instance plays in part upon the French proverb "Les yeux sont le miroir de l'âme" ("The eyes are the mirror of the soul"). The painting itself is of a large singular eye, the iris of which rather than being merely blue or green or brown is in fact full of clouds that surround the pupil and give the impression of being a view through a window, not into the soul, but into the sky. Although the image is ambiguous enough on its own, the title only adds further ambiguity. In one popular interpretation, it announces that the painting is in fact a dead end, not a mirror to the soul but a *False Mirror*, a counterfeit—something akin to *La trahison des images* (*The Treachery of Images*), Magritte's painting from later that same year of a pipe famously accompanied by the words "C'eci n'est pas une pipe." But if *Le Faux Miroir* announces that it is not a mirror to the soul, so too does this arresting image of the eye allude to something grander in force, power, and significance—which may or may not be false as well. Man Ray, who had the painting in his living room for many years, called it the "eye of sky" and noted that "for once, a picture sees as much as it is seen itself," thus likening "the dispassionate gaze of the eye" in Magritte's painting to what Astrit Schmidt-Burkhardt calls "the totalitarian language of the eye of God."[35] Following this line of thought, *Le Faux Miroir*, in addition to alluding to the soul, also recalls the notion of the all-seeing divine. The fact that the image vacillates between allusions to the soul and to God himself is simply a reminder that, at least within Western traditions, the former always presupposes the latter.

Lyon clearly opts for an interpretation that understands the "falsity" in the painting's title to be self-referential and a metareflection of the limitations of artistic expression as a medium. By placing a reproduction of *La Faux Miroir* directly above the title of his book, he thus cleverly implies that "the electronic eye" of surveillance is a counterfeit as well—a counterfeit of the soul and of God. This implication finds support not only in the proclamation of faith in the book's preface, where Lyon announces his commitment to "Christian social thought," but also in the opening chapter, which is entitled "Body, Soul and Credit Card." That chapter plots the eclipse of the "soul" by identities comprised of the "coded sequences of numbers and letters"[36] that are the distilled bits of information about us, surreptitiously collected, fed into data banks, and processed by computers. The debate here is not whether we should be alarmed by the surreptitious collection and processing of our personal data by anonymous entities and corporations. On that point, we all share similar alarm and concern. Rather the debate is about what we gain or lose by positioning those dubious processes in a binary opposition with the Christian notion of an individual soul. To state this debate another way—or, better yet, to frame it within the undercurrents of Magritte's visual discourse—the debate is about whether the electronic eye has closed the window to the soul or, alternatively, whether the notions of the soul and God put clouds in our eyes, so to speak, and give us a false image not only of the world around us but also of surveillance in particular.

At some level, this is a debate that runs the gamut of the entire history of surveillance. While the presence of *Le Faux Miroir* on the cover of Lyon's book may channel Magritte's visual playfulness into a subtle critique of surveillance's electronic eye, so too does the image position Lyon's book in the middle of a struggle over the eye's symbolic significance. Jessica Riskin reminds us that "because of the intimate connection that philosophers and physiologists identified between vision and consciousness itself, the eye" has long been considered to be the site where "the drama of the mind-body split most forcefully took place," a drama that as Riskin points out concerns the notion of "the very soul itself."[37] That dramatic struggle dates at least as far back as the fifteenth century, when the artist Hieronymus Bosch "inscribed the traditional symbol for divine omnipresence and omnipotence in the pure form of a circle" and thus created "a lasting and impressive symbol for the unceasingly watchful eye of God."[38] Yet by the late eighteenth century, that symbol was blurred with a kind of double vision. The French revolutionists, in what Schmidt-Burkhardt describes as "a blasphemous reversal of Christian symbolism," appropriated the image, recast it as "the supreme eye of reason" and realigned its symbolic significance with the vigilant watchfulness of the revolution and by extension of the modern secular state as well.[39] Carefully documenting this evolution's relation to the emergence of surveillance technologies, Schmidt-Burkhardt characterizes that emergence as a shift from the notion of a divine God "as the paradigmatic world observer" to something akin to the idea of deus ex machina[40]—a characterization that thus charts the evolution of the image into new symbolic forms while simultaneously acknowledging that the symbolic traces of the divine were never completely expunged.

In many respects, Schmidt-Burkhardt's characterization makes sense, and her allusion to "the God in the machine" finds a ready compliment in the recent work of performance studies scholars like Jason Farman, who plots a similar genealogy for the concept of virtual reality. In his book *Mobile Interface Theory*, for example, Farman notes that "the term 'virtual' is drawn from the Latin word *virtus*," which, as he reminds us, "has had many meanings over the years including 'virtue,' 'force,' [and] 'power.'" Its most recent transformation, however, occurred in 1959 when the term "virtual" found new currency among computer programmers who started using it to characterize that which does not physically exist per se but which is "made by software to appear to do so from the point of view of the program or user."[41] But even this adaptation of the term in 1959 was not entirely novel, as Farman himself observes. The "use of the virtual," he argues, "as that which is not physical but emulates the physical (such as 'virtual memory' or 'virtual machine') gestures back to the 1600s, when people used it to refer to metaphysical ideas related to Christianity."[42] In fact, that same history of the "virtual" and the corresponding modern notion of "virtual reality" points toward one of the key conceptual areas where, according to Lewis MacLeod, the notions of divine omniscience and surveillance overlap: "Both derive authority," he notes, "through corporeal absence."[43] To follow the line of thought developed in the work of Schmidt-Burkhardt, Farman, and MacLeod into considerations of surveillance is thus to go down a path leading to a place where the conceptual framing of the very technologies now central to almost all surveillance systems—the technologies creating the virtual realities of surveillance in the public sphere—bear the discursive traces of Christian metaphysics: deus ex machina indeed!

Of Last(ing) Judgments: Interpellation and Social Sorting as Surveillance

Although Farman does not mention surveillance directly in the passages cited above, the conclusions that he draws from his observations about "the virtual" and its vacillation between Christian metaphysics and secular technologies are a key into what, following Louis Althusser's notions of interpellation and hailing, are arguably the foundational ideologies behind surveillance in general and the ways in which it constitutes us as subjects in particular. In this regard Farman's arguments have a lot to tell us. Building on the underlying metaphysical registers of the term "virtual," Farman concludes: "Connecting the virtual to the spiritual illuminates the historical idea that virtuality is non-physical and largely about potential or a process of becoming."[44] With regard to the question of surveillance, such conclusions point toward the ways in which surveillance systems impact directly on what potential exists for those who fall under its watchful eye. They influence who citizens of the surveillance society are perceived to be and who they are allowed to become. They regulate how individuals are identified and positioned politically within the public sphere. Ultimately, they determine available opportunities and hence also determine what roles individuals are allowed to play in society more generally. In short, at the site of the virtual—at the site where Christian metaphysics and secular technologies converge—surveillance always involves a moment of interpellation and hailing.

To speak of interpellation and hailing is to recall Althusser's most significant contribution to Marxist theory, his explication of ideology in the essay "Ideology and Ideological State Apparatuses." As is well known, Althusser's main argument in that essay challenges the conventional notion that ideology is a collection of ideas, a political philosophy or a worldview that one consciously espouses or advocates. On the contrary, for Althusser, ideology is dialectical. On the one hand, it constitutes us as subjects through the already existing and unavoidable structures and institutions—the apparatuses—of the social order that shape our interactions and, prior even to our birth, position us within political and economic hierarchies. Yet our own actions are part of that dialectic as well. They too are ideological, so argues Althusser, because they perpetuate those same structures and institutions. Acquiring the skill and "know-how" that are required to become a functioning member of society is an ideological act, Althusser maintains, through which individuals affirm, perpetuate, and submit to "the rules of the established order."[45] It is a process repeated in the countless daily acts that constitute—indeed, that "interpellate" or "hail"—individuals as subjects: acts, in other words, that identify, name, and situate individuals within the existing conditions of production and power relations.

For all its theoretical complexity, ideology in action is easy to identify, and Althusser suggests that the ideological processes of interpellation or hailing "can be imagined along the lines of the most commonplace everyday police (or other) hailing: 'Hey, you there!'"[46] Althusser imagines this moment of hailing as a scene on city streets, where, he argues, "the hailed individual will turn around" or presumably will respond in some like fashion because he recognizes "that the hail was 'really' addressed to him." In doing so, Althusser argues, the hailed individual "becomes a subject,"[47] affirming and simultaneously submitting to the authority of the call. Judith Butler has argued that this imaged scene is not nearly as simple or "dyadic"[48] as Althusser suggests, but before considering her critique in greater detail, I want draw attention first of all to Althusser's unspecified reference to "other" forms of hailing that are similar to but not the same as what he describes as "commonplace everyday police . . . hailing."[49]

With the advantage of hindsight, it is possible in this vague allusion to "other" forms of hailing to detect a slight prescience not only of the surveillance society to come but also of what is arguably David Lyon's major contribution to the field of surveillance studies, namely his numerous explorations of surveillance as "social sorting." Among the best of those explorations is his 2009 book *Identifying Citizens: ID Cards as Surveillance*, where he explores the multiple ways that ID cards are the tangible extension of "a sorting system that puts citizens into categories to be better 'seen' and thus differently treated by the state."[50] But in this regard, ID cards are merely an example. "New ID card systems are a species of surveillance," Lyon argues, "but they also share a key characteristic of much of contemporary surveillance in that they facilitate forms of 'social sorting.'"[51] From this point, there is an identifiable path between Althusser's arguments that interpellation and hailing reproduce the conditions of production, and Lyon's sense that social sorting tends to serve the interests of "political constituents and corporate shareholders."[52] The more interesting issue, however, is the foundation beneath that path: the willingness of individuals to answer the Althusserian hail or to acquiesce to the technologies of social sorting. Both assume a predisposition to submission.

Exploring the similarities between Althusser's notion of ideology as interpellation and Lyon's notion of surveillance as social sorting may come across as a bit of a digression, but the question of predispositions toward submission returns us to Butler's critique of Althusser and ultimately locates those similarities in the conceptual metaphors of Christian discourse once again. Butler is particularly fascinated with the street scene imagined by Althusser. With regard to the policeman and the individual whom he hails, Butler cannot help but wonder, "What kind of relation already binds these two such that the subject knows to turn?"[53] We might pose an almost identical question with regard to the processes of social sorting to which individuals acquiesce. Lyon has repeatedly noted social sorting's cultivation of "social stereotypes" and its use of "unrefined ethnic-racial type of categories," and he suggests in his book *Surveillance Studies: An Overview* that "the big question then becomes: who defines these categories?"[54] But at a more fundamental level, the question is why do we acquiesce to them? What kind of relation already binds the sorted to the categories such that the subject knows to submit to the sorting and to surveillance technologies more generally?

Perhaps the answers to these questions are to be found in considerations of the ways in which policing and surveillance technologies are simply manifestations of a society that despite its democratic pretenses is nonetheless regulated through the exercise of brute force, intimidation, and subtle forms of state-sanctioned domestic terror. I don't want to discount such considerations and in fact will return to them momentarily. But even if those considerations point in the right direction, we must consider how they are cloaked and masked: hence the concern with Althusser, with his theories of ideology and with Butler's critique. Taking up the question of why the pedestrian knows to turn in answer to the policeman's hail, Butler challenges Althusser's claim that the pedestrian's willingness to answer the hail "cannot be explained solely by 'guilt feelings,' despite larger numbers who 'have something on their conscience.'"[55] That "something," Butler argues, suggests that "a prior and unelaborated doctrine of conscience"[56] exists within and is necessary for the functional viability of Althusser's theory of interpellation—a doctrine that, she argues, is grounded in Althusser's overreliance on Christian discourse to illustrate his theory.

Nowhere is that reliance more pronounced than in the last major movement of Althusser's essay, which bears the subheading "An Example: The Christian Religious Ideology." As an added backdrop to the earlier authoritarian image of the police officer hailing the pedestrian, in this final section of his essay Althusser offers the more arresting image of God naming Peter and hailing him as a spokesman (like the apostle St. Peter himself) to tell us that "God exists and that . . . [we] are answerable to him."[57] Folding Althusser's reference to Peter back onto his earlier narrative of the police officer hailing the pedestrian, Butler thus argues that Althusser has essentially stacked the deck and predetermined the results of his theory. With the hailing of Peter looming in the background, Butler argues, "The authority of the 'voice' of ideology, the 'voice' of interpellation, is figured as a voice almost impossible to refuse."[58] In this respect, the answer to the question of why the pedestrian responds to the police officer's hail is ultimately metaphysical. As Butler notes, in Althusser's narrative of interpellation the secular "postulated voice of the representative of state authority . . . the policeman's voice" resonates with the authoritative voice of God.[59] Turning in response to the hail, the pedestrian may become "a subject," but that is because in Althusser's theory he or she was already prefigured as "a soul." This then is the consequence of conceptualizing a theory of ideology through the tropes of Christian discourse: the theory is geared toward a reaffirmation of the notion of the soul and the notion of its immortal peril.

If Althusser's discussion of God hailing Peter were an isolated example, perhaps it wouldn't matter so much, but from this example Althusser extrapolates a broader claim: that the foundational assumptions of Christianity are paradigmatic of all Western ideologies.[60] Two possibilities follow from this assertion. The first is that Althusser may simply have recognized the deep impact that Christian beliefs have had on the functioning structures of Western society. The second possibility—and this is what implicitly Butler is arguing—is that in the articulation of his theory Althusser succumbs to a fallacy of composition (often called the "synecdoche fallacy"). Mistaking the example for the whole, Althusser inadvertently inscribes Christianity across the entire theory of interpellation and thus shies away, among other things, from a theory capable of contextualizing Christianity within an analysis of the raw exercise of power and authority and control—an analysis that, I would suggest in passing, would open the door to studies of ideology and surveillance. What Althusser offers instead is the all too familiar image of the Christian God weighing matters of "eternal life or damnation" according "to the respect or contempt . . . [individuals] show to 'God's Commandments.'"[61] As Butler notes, Althusser's notion of interpellation follows the script of the Christian promise of the soul's redemption in that "that social existence, existence as a subject, can be purchased only through a guilty embrace of the law, where guilt guarantees the intervention of the law and, hence, the continuation of the subject's existence."[62] With regard to what Butler identifies as ideology's underlying "doctrine of conscience," the answer to the question of what binds the wayward pedestrian to the policeman is thus ultimately a doctrine that preaches the presumption of guilt, a presumption that we might characterize as a modern adaptation of the biblical notion of original sin. For it is not merely the police officer that presumes the guilt of the pedestrian. The pedestrian, in an affirmation of the tropes of Christian faith, presumes his or her own guilt in responding to the police officer's hail.

The convergence of Althusser's theory of interpellation and the discourse of Christianity bears a familiar refrain: it is akin to the initial image of *The Archbishop's Ceiling* with its Christian iconography and surveillance technology mixed together. It is akin to the eye of God blurring into the watchful eye of the revolution and of "the modern police state."[63] It is also akin to the concept of virtue traveling the path from Christian metaphysics to secular technologies and surveillance systems. Somewhere beneath that refrain, however, I want to suggest that what we discover is not just "the doctrine of conscience" but a very different kind of ideology of which the doctrine of conscience is merely a constitutive part: one that rationalizes intimidation, abuse, exploitation, and the use of brute force by normalizing a culture of suspicion and by granting individuals access to subjecthood, citizenship, and even basic human rights only to the extent that they accept that they are always already deserving of suspicion and are thus willing to submit to whatever passes as necessary regulation, control, and oversight. Whether they literally and consciously accept this is inconsequential, for when the policeman hails or when surveillance systems sort and interpellate, the wayward pedestrian always yields and, one way or another, heeds the call. But if Christian metaphysics and the doctrine of sinfulness contribute to this process by cultivating the receptivity to the notion that we are all deserving of suspicion, then perhaps by challenging the underlying assumptions of that receptivity—perhaps by confronting its Christian foundations head-on—we might expose the brute material realities for which they provide cover and confront them as well.

Butler's goal in her critique of Althusser is to disentangle the theory of interpellation from the unacknowledged "doctrine of conscience" that regulates it and thus to salvage the theory with a vision of ideology unobstructed by Christian metaphysics. In this regard, Butler's critique offers a powerful example, which I would suggest can and indeed needs to be extended to the theory of surveillance as well. Perhaps as an illustrative point of departure for such an extension, we might play image against image and counter Althusser's closing example of God hailing Peter as a spokesman with a demystification of the biblical narrative of Peter denying three times that he is one of Jesus's disciples when hailed by the high priest's officers and servants after Jesus's arrest. In the biblical myth of this narrative, Peter is answerable only to God for those denials, which were prophesized to him and which, with the cock's crow, constitute him as God's subject and do so in a moment of punctuated shame, guilt, and denial—thus reaffirming the doctrine of guilt and sin. In the modern era of surveillance technology and face recognition software, by contrast, Peter's denials would not have thwarted the confirmation of his identity and apprehension by the authorities, who would have rendered him a very different subject than that which he is in church history and lore. Rather than being granted the keys to the kingdom, he would have been placed under lock and key and then executed. His fate would have become an example of what to avoid; his execution would have been the example reinforcing a culture regulated through the exercise of brute force and intimidation.

What remains as the unanswered question at this Althusserian site is whether contradicting the doctrine of guilt might break the cycle of interpellation, dispel ideology's aura, and alter the exploitative conditions of production and power. A similar yet arguably more daunting question also looms over dominant conceptions of surveillance as well, particularly with their overreliance on tropes borrowed from Christian discourse, and in this regard, we find ourselves at the limits of Butler's critique of Althusser. Adam Kotsko has suggested that the major shortcoming of Butler's critique is a failure to recognize that Althusser's objective in his turn to Christian doctrine was not to "sacralize ideology" but rather to challenge both religion and ideology as illusions.[64] This, Kotsko notes, is a reading that is more consistent with the skepticism that as a Marxist Althusser had toward religion in general. But this is a skepticism not shared by all. Nor is it a skepticism that can be taken for granted in the field of surveillance studies, which unlike Marxist philosophy is not positioned in opposition—or as an alternative—to the discourses of Christianity, its notion of the soul, or its doctrine of conscience.

Omniscience's Ruse: The Compartmentalization of Surveillance in Bentham's Panopticon

On this last point, we return once again to David Lyon's *Christians & Sociology* and to his dismissiveness there of scholars who would argue "that people do not have a 'self' but rather a succession of roles."[65] Even though I would argue that this rather categorical and reductive dismissiveness of the theories postulated by his fellow sociologists was as misguided in 1975 as it is today, the structure of that dismissiveness is consistent with the underlying assumptions of Lyon's critique of surveillance. If the effect of surveillance is to determine who individuals are, who they become, and what roles they are allowed to play in society, then Lyon's concern with surveillance might well be characterized as an extension of his earlier concerns in *Christians & Sociology* and hence as a defense of the "self" against its diffusion within the technologies of the surveillance society. There is an echo in this defense of the Christian sociologist Jacques Ellul and his much-celebrated 1954 work *The Technological Society*. There, Ellul famously warned of the threat that the growing dominance of technology ("la technique") poses to our humanity,[66] and, much to his credit, at some level Ellul certainly foresaw—at least a generation before the rest of us—the kind of society in which surveillance is now an integral part. The more important issue at hand, however, is that as points of opposition to the broadly defined notion of a technological society or to the more specific notion of a surveillance society, Ellul's sense of humanity and Lyon's notion of the self share much common ground. Both, I would suggest, are euphemisms for the Christian notion of the soul—a notion that presupposes not only the idea of guilt and sin but also the watchful eye of God. In fact, Lyon's early critiques of surveillance have that watchful eye in mind, and his implicit defense of the self or the soul involves a strategy of positioning the eye of God rhetorically as a contrast to—indeed as an antidote for—what he characterizes as the suspect pursuit of "secular omniscience."[67]

The problem is—and it is a problem that calls for careful deconstruction—the more Lyon summons Christian discourse as a counter to surveillance, the more his work becomes entangled in an affirmation of the conceptual metaphors that from its inception have facilitated the rise of surveillance itself. As I mentioned earlier in this chapter, the sleight of hand in that discourse—its pivotal binary opposition—is to present God's omniscience as force of care while contrasting it with an image of surveillance as a secular force of coercion and control. And yet, as was evident from our earlier discussion of *The Abbess of Crewe*, the two ultimately bleed into one another, and not just in works of fiction. That blurring is also evident in Lyon's handling of the impact that the image of an all-seeing God had on the late eighteenth-century secular reformist Jeremy Bentham, who capitalized on that image of God as an apt metaphor for the architectural technologies utilized in what for many is the paradigmatic touchstone of surveillance itself, namely his model of the panopticon.

Just as the French revolutionists usurped the image of the watchful eye of God for their own embrace of secular reason and for the watchfulness of the Revolution, so too did Bentham's original proposal for the panopticon borrow heavily, as Lyon notes, from "religious sources"—like the passage from 139th Psalm that served as its epigraph: "Thou art about my path, and about my bed: and spiest out all my ways." Indeed, the proposal "drew freely on . . . [Bentham's] understanding of God's omniscience."[68] But rather than seeing Bentham's use of these sources as a legitimate moment of secularizing them and of developing them as valuable conceptual metaphors to illustrate a new sociopolitical context, Lyon adopts a more orthodox mode of exegesis. In response to Bentham, he chooses to police a strict notion of scriptural authority and meaning—one that, presuming divine inspiration, leaves little room for the multiple meanings that a text may take on in different sociohistorical, cultural, or political contexts.

The irony of this response is that Lyon's critique of Bentham is thus premised upon a strict fundamentalist reading of scriptural intent that slides into its own kind of authoritarianism, a reading that disciplines Bentham and those who would follow him for not appropriately acknowledging the benevolence of God's omniscience. For example, Lyon dismisses Bentham's panopticon as "a secular parody of divine omniscience"[69] because it does not affirm the Christian notion that the eye of God "protects and enables," and watches over us with "loving care" in a manner that is "personal, relational, and mutual."[70] Recently, these same arguments have been recycled almost verbatim by Eric Stoddart, who like Lyon before him, argues that the classic model of surveillance found in Bentham's panopticon is little more than "a parody of God's omniscient gaze" since it neither emphasizes a notion of God nor of surveillance itself that culminates in an affirmation of benevolent "Divine care."[71] Yet it is never clear in Lyon's or Stoddart's critiques how Bentham's adaptation of the notion of God's omniscience counts as parody simply because it doesn't reaffirm the notion of a loving God.

Broadly speaking, parody is defined by what it does rather than does not do, and what parody does is to subvert authority in a generally playful manner. It is more suitable to what, in his book *Jesus in Disneyland*, Lyon cites as examples of the intersection of religious belief and postmodern sensibilities where parody is an integral part of a cultural and political aesthetic. In his book, Lyon alludes to the notion of going to "cyberchurch" as an example of this intersection.[72] But perhaps the more significant discussion of "cyberchurch" is to be found in Mark Dery's wildly entertaining 1996 book *Escape Velocity*, which contextualizes the notion of "cyberchurch" within the consciously irreverent cyberpunk and technopagan subcultures of the 1990s. As Dery notes, those subcultures achieved a noteworthy presence within literary and artistic circles, in large part by playfully blending a kind of "neopaganism" and New Age religious philosophy "with digital technology and fringe computer culture."[73] The result was a consciously subversive (if somewhat short-lived) counterculture that was fueled by its calculated parodies and "hacking" of mainstream values and beliefs. This was parody, and unlike Bentham's proposed panopticon, it sought to subvert rather than to find legitimization within the mainstream. Rather than seeking parody, Bentham's appropriation of the tropes and metaphors of religion was more than anything else a quest for legitimacy.

With or without metaphors borrowed from Christian discourse, there is nothing particularly funny or amusing about the panopticon.

The panopticon is a grim place, where, as Michel Foucault has famously noted, the "perfection of power" is exemplified in the grotesque fact that inmates are "caught up in a power situation of which they are themselves the bearers."[74] If anything, it is the lack of parody that ought to be alarming in Bentham's model, and we could do far worse than to ask ourselves how we became so calloused as to not be horrified by the psychological torture that the panoptic model applies to inmates. The panopticon was not conceptualized to ridicule Christian beliefs through humor or parody but rather to capitalize on them in a new context and to take advantage of the way that internalizing the belief—rightly or wrongly—that one is being watched over presumably affects consciousness and hence behavior. In this respect, it doesn't matter whether the internalized image is that of a vengeful or benevolent force. What matters is that one submits to the authority symbolized by the internalized image and that one modifies one's behavior accordingly. With the threat of eternal damnation or of indefinite incarceration looming as a consequence of a lack of submission, conformity in this regard is always already a product of duress and threat. One can call it God's tough love or whatever one wants, but it is still an exercise in coercion and intimidation.

While there is no need to hold Lyon and Stoddart to a strict definition of parody, it is important to consider what lies beneath their use of the term and what this meaning indicates for how they would frame our understanding of surveillance. It is not too difficult a step to recognize that "parody" for Lyon means an unauthorized departure from orthodox doctrine: a departure that presumably makes a mockery of what the doctrine stands for. In the broadest sense, Bentham's offense is to have rearticulated the sacred as secular, to have transcribed the spiritual into the material, and thus to have made the holy profane. But on a smaller scale, his selective sampling of Christian doctrine is the stuff of heresy—at least among the faithful—and far from responding with an outreach of benevolence to heretics, the church has a long history in the face of heresy of resorting to discipline, often quite publicly and severely punishing the heretic so as to close ranks and maintain doctrinal integrity, purity, and authority.

On a small rhetorical scale, a similar reaction surfaces in the concluding paragraph of Lyon's essay on Bentham. Alluding to Bentham's appropriation of Christian discourse, Lyon accuses him of secular utopianism and warns about his misuse of scripture: "We forget its origins at our own expense."[75] At our own expense? What price Lyon means is never stated, but the implication, while equivocal, is hard to miss. Lyon comes full circle with this conclusion. Beneath the guise of asserting the enabling "loving care" of God's omniscience, Lyon indirectly reminds his readers of God's authority and of the price to be paid for not adhering to it: that ever-looming specter of eternal damnation. I want to suggest that at this moment of what amounts to a metaphysical threat, the metaphors employed by Lyon lose much of their distinction from those employed by Bentham. Both rely on a cultivated belief in the presence of a punitive watchful eye, which may or may not be present but which nonetheless exercises ultimate authority and control. But more important still, both rely on an accompanying rhetoric that smooths over their authoritarian undercurrents. The rhetoric may pay lip service to benevolent care or to public security, but either way it sugarcoats the implicit threat.

Bentham's sin was to have made this threat visible and tangible. It was to have shifted the metaphysical threat to physical, material reality. Indeed, for all the scholarly excitement around Foucault's reading of Bentham, this is arguably the crucial point that Foucault missed about the panopticon as a model of surveillance—a point that Lyon misses here as well, largely as a consequence of being preoccupied with a struggle over secular representations of God's ostensibly benevolent watchful eye. In a notable departure from Foucault, Miran Božovič argues, for example, that Bentham's panopticon project was less concerned with inducing in the inmates what Foucault describes as "a state of conscious and permanent visibility that assures the automatic functioning of power"[76] than he was with staging the inmates' punishment to a public who would have access to the central tower of the panopticon. "Punishment" as "an example," Bentham wrote, is "the principal object."[77] The stunning implications of that principal focus are what is lost in a critique that would dismiss Bentham's project as "a parody of God's omniscient gaze." Rather than simulating the watchful eye of God in a manner that conceivably could be extended to the whole of society—as is the image conveyed in the biblical notion of God's omniscience—the objective, as Božovič notes, was deterrence, specifically deterrence that was the result not of being subjected to surveillance oneself but of seeing others subjected to that fate. Witnessing the terrifying wretchedness of the prisoners' situation staged before them, the public would likely be deterred from criminal activity, or such at least was Bentham's reasoning.[78]

As a social concept, deterrence is easy to take for granted, so it is important to be clear about its meaning. It is a euphemism for intimidation and coercion through the threat of force; and in Bentham's proposal, for a threat by way of visible example. Beneath the discourse of what Božovič calls Bentham's "fiction of God," the actual functioning mechanism of surveillance as conceptualized through the model of the panopticon was limited to the panoptic prison itself—where through surveillance inmates were made examples of—so as to cultivate social awareness of how horrible and wretched it is to be ensnared within a calculated punitive system of seeming omniscience. In this respect, panoptic surveillance in Bentham's model was never a prescription for the citizenry as a whole. It was a form of punishment for the few, and "All punishment in itself," Bentham wrote, "is evil."[79] Like Bentham, one may debate whether there are necessary lesser evils in society that prevent greater ones. This is in fact the logic behind current justifications for the continued expansion of surveillance technologies deeper and deeper into our private lives. But following Bentham's lead in this debate, perhaps a good point of departure for any discussion of surveillance would be to recognize that surveillance is first and foremost—always already—a punitive measure, and that when all is said and done, biblical metaphors tend to obscure this basic historical fact.

We can cloud our eyes with images cribbed from biblical notions of God's benevolent watchfulness and do so in such a way that would cast surveillance as a force that also has a benign side, providing oversight, security, and care. But those notions perpetuate—and indeed, are arguably the source of—the contradictions that haunt Bentham's proposed panopticon, which while offering, on the one hand, what it implied was a more humane and compassionate form of incarceration, ultimately proposed, on the other, to parade unknowing inmates before the public so as to frighten citizens into social conformity with the threat of a similar horrific fate. At least within Bentham's proposal, surveillance was contained within the walls of the panopticon and identified clearly as an instrument of punishment. Today, that instrument has been repackaged and remarketed, and it pushes beyond the prison walls toward ubiquity and omnipresence. But we ought not delude ourselves about what remains: its fundamental punitive character and its primary use as a mechanism of intimidation. The difference today is that rather than parading inmates before the public as examples of a fate to be avoided, the strategy is to give everyone a taste of the punitive: a little punishment to frighten them with the prospect that there might be more to come. This is a strategy, I would suggest, that is only palatable to those inclined to believe that we are all guilty and sinners anyway—to those, in short, whose eyes are clouded with what Butler calls "the doctrine of conscience" and who believe punishment is what we deserve.

Ratzinger's Letter: Of *Doubt*, Acceptable Risks, and a Culture of Abuse

While this latter formulation about clouded vision implicitly returns us to Magritte, it is more than a passing acknowledgment of a plausible alternative interpretation of *Le Faux Miroir*. Inasmuch as it refers to the effect of religious discourse on our understanding of surveillance, it also includes an obvious adaptation of Marx's own critique of religion, which is best known not for the equation that the critique of religion is the prerequisite to all critiques, but for the more famous assertion that follows two paragraphs later in the same essay. "Religion," Marx argues there, "is the sigh of the oppressed creature, the heart of a heartless world, and the soul of soulless conditions. It is the opium of the people."[80] One need not subscribe entirely to Marx's somewhat one-dimensional characterization of religion as a ruse or, to echo Magritte once again, as a kind of false mirror designed to distract the oppressed from the reality of their material conditions. However, there is, as I have been arguing throughout this chapter, still room for a legitimate embrace of that characterization as a cue for giving serious attention to the ideological effects of Christian beliefs on attitudes toward surveillance.

Keeping this same cue in mind, I would suggest that Marx's assertion resonates well with works like John Patrick Shanley's Pulitzer Prize–winning play *Doubt*, which explores the ethical complications that result from a nun acting on her suspicion that a parish priest by the name of Father Flynn may have molested a young African American boy at the Catholic school where she is headmistress. Although the hierarchies of the church necessitate that she proceed with calculated discretion, Sister Aloysius employs the limited authority that she has as headmistress not only to put Flynn under watch but also to reconcile her orchestrated activities with the mission of the school to provide "constant educational, spiritual and human vigilance."[81] Indeed, she effectively co-opts her newest teacher, Sister James, into the surveillance of Flynn, first by telling her that the students need "a fierce moral guardian" and that if she "is vigilant, they will not need to be."[82] Laying this foundation, she then directs Sister James toward vague "matters in St. Nicolas School" and ultimately toward Flynn himself.[83]

Although Sister Aloysius proceeds against Flynn with fervor and conviction, the play never resolves the question of whether the headmistress's suspicions were actually correct, and, at one level, what makes the piece so compelling is the unresolved question of whether Flynn is in fact a pedophile priest or alternatively a closeted homosexual who is the victim of the gross stereotype that equates homosexuality with pedophilia. Elizabeth Cullingford argues, for example, the play allows for the possibility that "Flynn could have been the closeted priest" that Sister Aloysius "thought he was without being a child abuser" and that Flynn's relationship to the student (whose mother identifies him as homosexual) "might have provided a vulnerable gay child with invaluable positive mentoring from a sympathetic gay adult."[84] All of these possibilities cultivate the notion of doubt that haunts the play's ending and that highlights the extent to which Sister Aloysius loses her way in the fervent pursuit of her convictions. As Cullingford rightly observes, at some point Sister Aloysius begins to care "more about her crusade against Father Flynn than about the vulnerable boy in whose defense she started that crusade."[85] Not only does this lead her into ethically questionable practices like trying to coerce the boy's mother into cooperation by threatening to expel her son from the school, but it also leads her to rationalize dubious acts like lying to Flynn about having contacted his previous parish and having found other nuns who will speak out against him.

There is much in Shanley's play that figures into a general critique of the institutions of Christianity. But the most important aspects of *Doubt* as a work that might be cited with regard to the intersection of Christianity and surveillance concern the fact that the play's most suspect—indeed most *dubious*—moments of surveillance are inextricable from its characters' deep religious convictions or from the social climate created by those convictions. They are not the counterweight to the electronic eye of surveillance. They are ultimately its rationalization—even if the instances of "surveillance" in the play are rather rudimentary in form. Here, *Doubt* pushes beyond the margins of its own fictional conceits and into the history of the church itself. For in the play, and arguably in the history of the church as well, the assumption that all are sinful, guilty—and hence in need of oversight, shepherding, and care to keep them on the path of the righteous—provides a justification for profoundly suspect invasions of privacy.

Such assumptions also provide a rationale for surveillance regimes more generally and are arguably a plausible source of the casual attitude of those who claim that the proliferation of surveillance technologies does not alarm them because they have nothing to hide. At some level, Shanley implies precisely such a connection in *Doubt* when Sister Aloysius tells her newest teacher, Sister James, that the students "should think that you see right through them" and that "only the ones that are up to no good" will be frightened by the intrusiveness.[86] She applies this same categorical logic to Father Flynn when she lies to him about having contacted a nun at his previous parish, and concludes from his subsequent resignation that it "was his confession" and that "he was what . . . [she] thought he was"[87]—even though his resignation might very well have been the product of anxiety about bigoted enemies at his old parish who, for whatever reason, would be as willing to lie just as Sister Aloysius has so as to do him in. In this scenario, Flynn opts for privacy and a transfer over a public fight in order to maintain some sense of human dignity.

Shanley's play lends itself equally to the opposite reading of Flynn, of course, and that is what gives the piece its staying power. But with regard to surveillance, the point to be drawn from Shanley's *Doubt* is that what surveillance seeks and what it exposes are not inherently in accord. Similarly, what surveillance finds is not necessarily what its operative logic suggests it has found. What is of no legitimate concern to the operations of surveillance—what one holds in discretion and privacy—may be that which surveillance discloses and spins in unpredictable directions. In surveillance, there is no objectivity of fact or truth, as is presumed in Christian doctrine. Surveillance is all too human, and human bias underlies the assumptions that filter and frame what it seeks, and human bias underlies assumptions governing the interpretation of what it claims to have discovered. Neither necessarily produces reliable results. Surveillance is, in other words, a regime for manufacturing truths, the validity of which are matters of faith. Indeed, with regard to surveillance, one could do far worse than to heed the closing lines of Shanley's play and echo Sister Aloysius's confession: "I have doubts! I have such doubts!"[88]

At the very least, those closing lines might provide a healthy dose of skepticism vis-à-vis arguments from theologians like Eric Stoddart who, in what could easily be taken as a doctrine borrowed from Sister Aloysius, has suggested that the very notion of privacy might be indefensible from a Christian perspective. In his book *Theological Perspectives on a Surveillance Society*, for example, Stoddart argues that "there might be no legitimate notion of privacy for anyone, given that God knows everything about us all." Following this line of thought, he concludes, "Christians, in support of privacy, are attempting to defend the indefensible."[89] Such conclusions do not counter the increasing encroachments of surveillance into our daily lives. Nor do they advocate resistance to such encroachments. On the contrary, Stoddart's conclusions are simply novel articulations of a well-established church doctrine that would have the faithful not only look beyond present circumstances but also look to their church representatives and submit to their authority, trusting those who exercise pastoral oversight to shepherd them toward a future better place. One can legitimately debate the merits of this doctrine. But the point is that Stoddart is talking about a lot more than the authority of the church and its representatives.

Casting a theological aura about surveillance, Stoddart's book admonishes us to have confidence in its regimes based upon the notion that those who are granted authority can and often do act with care and compassion, regardless of whether they are figures of the church, agents of the state, or nebulous corporate entities. The operative assumption here combines a reaffirmation of Christian notions of compassion with an endorsement of the care and oversight that state institutions and corporate organizations ostensibly provide. From an ideological perspective, the message to those who fall under the purview of the church, the state, or other organizations is equally clear. Those who believe that we are by nature sinful—those who believe that god's omniscience renders the notion of privacy indefensible—are those who are generally predisposed not only to believe in the necessity of shepherding and surveillance but also in the fundamental affinity between the two and hence in the virtue of submitting to both. Rather than dispelling this notion, Stoddart actively encourages it, and would have his readers frame their understanding of surveillance within the "perspective" of a generalized notion of Christian "care." Indeed, he likens surveillance to the responsibilities of Christian "parenthood, education, health and even friendship."[90] Beneath these suggestions, there is the larger assumption that Stoddart's theological approach to surveillance encourages: the assumption that surveillance's ideal potential should be the focus of our attention rather than its current flawed manifestations.

Dismissing concerns about privacy, Stoddart encourages a belief in the possibility of something much more benevolent in the future: the possibility that surveillance can be more than it actually is and the possibility that somewhere in the future one can "practice surveillance compassionately."[91] There is an echo in this encouragement of Lyon's own arguments that "surveillance is not unambiguously good or bad,"[92] and of his argument that given surveillance's "positive social benefits" we might legitimately question the need to "resist systems whose advantages simply carry with them a number of acceptable risks."[93] Lyon has since become increasingly critical of this seemingly neutral stance. But even in more recent publications like his essay "Resisting Surveillance," where he argues that "treating ordinary citizens as suspects . . . is an unlikely recipe for confidence in the political process,"[94] he nonetheless returns, as an alternative model, to a notion of "security" derived in large part from arguments that Jacques Ellul, the author of *The Technological Society*, formulated in his later 1970 book, *The Meaning of the City*. This later book charts a biblical journey from Babylon to the New Jerusalem, and through the work of Ellul, Lyon thus subtly embraces the Christian image of the New Jerusalem where security is the product of God's eye watching over us all: there surveillance is indistinguishable from benevolent care. But as we all know, the Jerusalem of the here and now is nothing like this. It is a divided city immersed in the technologies of surveillance, military protocols, ethnic conflict, and antagonistic religious doctrines.

Something all too familiar rings in the logic of these arguments from Stoddart and Lyon. They are embedded in the common doctrinal refrain of the present being deferred in favor of a promised future or a promise land. This is the doctrine that Marx himself argued against in the introduction to his "Critique of Hegel's 'Philosophy of Right.'" It is the doctrine that he likened to the effects of an opiate. Only in this instance, the doctrine that distracts believers from the material realities of the here and now with an otherworldly promise is the same doctrine that provides ideological cover for an ugly and dangerous shadow that plagues the surveillance society—a shadow that, I would suggest, it always produces and cannot escape—the shadow of unchecked opportunity for abuse. I will talk at greater length about that shadow in the third chapter of this book that addresses prisons as maximum surveillance environments. But for now suffice it to say that if the doctrines of the church provide ideological cover for abuse within the structures of the surveillance society, that is because they produce and hide similar opportunities with the structures of the church itself.

Granted there are the larger material conditions that are buried by the otherworldly focus of Christian discourse. But at a more basic level, that discourse mystifies and obscures the internal workings of the church, and thus beneath the cloak of pastoral oversight and care it also cultivates a climate ripe for abuse. The parallels here are striking. On the one hand, a doctrine that questions the need to "resist systems whose advantages simply carry with them a number of acceptable risks" is an example of the kind of special pleading that rationalizes away surveillance's darker probabilities and its questionable inevitabilities. On the other, it is also the kind of doctrine that translates back, for example, into a defense of a church confronted with mounting revelations not merely of sexual abuse by members of its clergy, as is suggested in Shanley's play *Doubt*, but also of cover-ups of those abuses by the members of the church hierarchy—cover-ups that at some misguided level discount abuse as an "acceptable risk" within the larger ministry that the church provides. The examples here are almost too numerous to count, but one stands out in particular.

Perhaps the most egregious example of doctrine facilitating church-sanctioned cover-ups is to be found in a leaked 2001 letter penned by then-cardinal Ratzinger that was sent to every Catholic bishop. Elias Groll has noted that Ratzinger in his role as a cardinal and later as Pope Benedict consistently "displayed a preference . . . for confronting predatory priests behind closed doors and [for] protecting the church's reputation at the expense of public accountability."[95] In his letter to the bishops, Ratzinger not only asserted "that the church" rather than "civil authorities"[96] has juridical authority in matters pertaining to sexual abuse perpetrated by priests, but he also informed the bishops that the church had the "right to hold its inquiries behind closed doors and keep the evidence confidential for up to 10 years after the victims" turned eighteen and became adults. (In practical terms, this means that the family of a child abused at the age of eight would have to wait twenty years for access to the evidence presented in the church tribunal.) In the conclusion of his letter, Ratzinger sternly reminded the bishops that "cases of this kind are subject to the pontifical secret" and threatened anyone breaching that secret with excommunication.[97] With the threat of excommunication, cover-up had the blessing of the church as official church doctrine. One might suggest that this does not sound at all Christian, but at some level such a suggestion is special pleading: it is an ideological act of trying to preserve the reputation of a privileged discourse from the material realities of what it ultimately facilitates.

Ironically, the special pleading of such a suggestion is, in fact, consistent with the logic employed by Ratzinger himself in his letter to the bishops. In the name of protecting the church's greater good—in the name of preserving the reputation of the church's privileged discourse—he helped to bury the church's more compromising realities from public view somehow without realizing that the logic he used to rationalize church-sanctioned cover-ups was part of what created the problem in the first place. That problem takes us directly back to the image of "the overpowering and ubiquitous eye of God," which Schmidt-Burkhardt associates not with the Christian notions of compassion and care but with the privilege of "free vision" (i.e., vision not limited or restricted) that is available only to a select few and that thus serves, she rightly argues, as the quintessential prototype of "a hierarchical system of authority."[98] But where does this leave us with regard to surveillance?

At this juncture, one important conclusion is certainly within grasp. Beneath the discourse of God's omniscience—beneath theological arguments that, from a Christian perspective, privacy is indefensible—the sociopolitical structures of oversight, authority, and surveillance cultivate and sustain political hierarchies that thrive at the expense of transparency. This is as true within the discursive economies of the church as it is within the discursive economies of surveillance. Indeed, as a sociopolitical phenomenon surveillance is always purchased at the expense of transparency, and inasmuch as Christian discourse embraces the notion of God's omniscience, it provides the ideological currency for the transaction. But there are more conclusions to be drawn. Perhaps the most disturbing among them is the probability that surveillance lends itself so well to Christian metaphor because both are fundamentally hierarchical in structure and effect, and inasmuch as both endow those hierarchies with authority that defies transparency, they foster a culture of abuse. In this regard, the workings of the church rather than its doctrines offer profound lessons of relevance to the study of surveillance.

Amid the ultimate revelation of an almost incomprehensible number of abuses by priests around the globe, who were as likely to prey upon the flock as to pray for them, it is difficult to know which is the greater offense, the serial abuses by predatory priests like a Peter Hullermann in Germany and a Lawrence Murphy in the United States, or the secretive machinations of the church hierarchy: the bishops, archbishops, cardinals, and a future pope who facilitated and compounded those abuses by using the authority of their offices to shuffle predatory priests from one perish to the next, covering up their crimes either in convoluted moral gestures aimed at preserving the reputation and ministry of the church—the very reputation that their own actions undermined—or in deeply misguided and selective acts of compassion for the perpetrators rather than for the victims.

Cite as many cases as we may, there is no need to rehearse the lurid details of these heinous acts in order to understand the pertinence that they have to the discussion of surveillance. For if the critique of religion is to serve as a prerequisite to the critique of surveillance, then one needs to look no further than to the church's empowered apologists: those who in the name of the faith's ostensible higher aims would rather admonish the faithful to see abuse as an aberration and misuse of sacred doctrine than encourage them to consider the ways in which the actions of predatory priests and their like are in fact examples of the kind of acts that the doctrine of the church accommodates. This is not to suggest that the church produces pedophiles. It is rather to suggest that its structures and discourses will always attract those with predatory inclinations because they provide them not only with ample opportunity to act upon their inclinations but also to act with the aura of absolution and impunity.

The church is certainly not unique in the attractiveness that it has for those of predatory nature. As a sociopolitical form, surveillance has a similar structure and is comparably attractive to those inclined toward corruption, exploitation, and abuse. But the crucial difference here is that unlike the discourses of religion, those of surveillance need not have faith as their cornerstone, particularly the New Testament variety, which defines faith as "the substance of things hoped for, the evidence of things unseen."[99] It is the logic of that definition that underlies Lyon's notion of "acceptable risks": a willingness to turn a blind eye to abuses that are manifest in order to maintain a belief in things hoped for and unseen. But what Lyon presents as "acceptable risks" ought to be viewed as inevitable outcomes that will continue to expand. Rather than being built from a cornerstone of things hoped for and a belief in things not yet manifest or seen, surveillance must be assessed empirically. Our attitudes toward surveillance must be grounded in a skepticism profound and strong enough to question what surveillance might bring in light of the abuses it already accommodates. As simple as this suggestion might sound, it is in fact a call for a radical departure from the status quo, and in practical terms, it means a fundamental reorientation of our critical approaches to surveillance. It is, in short, a call for a critique of surveillance that has fulfilled the prerequisite of having first provided a critique of religion.

Surveillance Operatives: Activists, Undercover Agents, and the "New Theater"

> [The] purpose of counterintelligence action is to disrupt . . . and it is immaterial whether facts exist to substantiate the charge.
>
> —J. Edgar Hoover

Of Garbage Cans, Undercover Operatives, and the Alchemies of the "New Theater"

In 1958, the 1960s were already in sight, and if technically they had not already begun, conceptually at least they had—particularly with regard to the radical redefinition of theater in the hands of the unconventional collection of artists whom Michael Kirby would later group together in the introduction to his now classic anthology *Happenings* (Dutton, 1965). It was in the introduction to that anthology as well as in the essay entitled "The New Theatre," which he published in the same year as *Happenings*, that Kirby first introduced the term "nonmatrixed performance," and spoke of how happenings, "as well as productions in other performance-oriented fields, force us to examine theatre in a new light and raise questions about the meaning of the word 'theatre' itself."[1] But even before Kirby raised this crucial self-reflective question of 1960s about "the meaning of the word 'theatre'" and hence about the meaning of theater itself, Allan Kaprow had already laid the conceptual groundwork for many of Kirby's arguments. Indeed, Kaprow was the central figure in Kirby's conception of "the New Theatre." More important, however, it was Kaprow who in his 1958 reflection "The Legacy of Jackson Pollock" suggested that after Pollock and abstract expressionism, artists were now "at the point where . . . [they] must become preoccupied with and even dazzled by the space and objects of . . . everyday life," and where they must now utilize "the specific substances of sight, sound, movements, people, odors, [and] touch." In this new blurring of art and life, according to Kaprow, the "present generation of artists," who were coming of age just as the 1960s were about to begin, would not only show us "the world" that we "have always had about us but ignored, but they . . . [would also] disclose entirely unheard-of happenings and events, found in garbage cans, police files, [and] hotel lobbies" and "seen in store windows and on the streets." Such revelations, Kaprow concludes, may leave people "delighted or horrified." They will leave critics "confused or amused," but either way, so argues Kaprow, they "will be the alchemies of the 1960s."[2]

Amid the garbage cans, hotel lobbies, and gritty city streets, it is easy to succumb to what, back in the mid-1980s, Barbara Haskell insightfully called "the aesthetics of junk," which carried artists like Robert Rauschenberg, Allan Kaprow, Jim Dine, Red Grooms, and Claes Oldenberg not only off the painterly canvas and into uncharted performance environments but also out of the abstract expressionism of the late 1940s and early 1950s and into "the new vernacular realism"[3] of the minimalism and pop art of the 1960s. In 1958, Kaprow apparently had a keen eye for what the impact of these aesthetics would be. People, on the other hand, were genuinely divided: both horrified and somehow elated with the implicit aesthetic glorification of their own junk. They were so horrified, in fact, that in the same year that Kaprow published his reflections on Pollock's legacy, *Newsweek* sounded an alarm against the "Trend to 'Anti-art,'" and labeled Kaprow "one of the dangerous radicals on the art horizon whose work threatened to overthrow sacrosanct values."[4] Haskell notes the irony of this sounding of alarm, given the fact that it was almost immediately followed by a resounding chorus of accolades. "A year after his appearance in *Newsweek*," Haskell notes, "Kaprow was selected as one of the up-and-coming artists in the prestigious Art in America New Talent Awards."[5] As much as Kaprow deserved the accolades he received, as much as he and his fellow artists deserved the recognition that Michael Kirby and *TDR* gave them in the mid-1960s, and as much as they earned the retrospectives given them by Barbara Haskell and the Whitney Museum in the mid-1980s, or by reenactments of happenings like that staged by Andre Lepeke in 2006, Kaprow and his colleagues were not the only workers of alchemy in the 1960s. They were not the only ones who could find gold in the stuff found "in garbage cans, police files, [and] hotel lobbies" or "seen in store windows and on the streets."

It is easy to overlook how much both of what Kaprow suggested his "generation of artists" would show us and of where he said it could be found reads like a briefing on undercover surveillance to a group of newly minted FBI cadets. Paying attention to what we "have always had about us but ignored," to what can be found in "police files" or in "trash cans," to what can be observed in "hotel lobbies," on city "streets," and in the reflections of "store windows," to what can be garnered from the "substances of sight, sound, movements, [and] people"—this is literally the stuff of FBI surveillance. Emily Roxworthy recalls, for example, that J. Edgar Hoover, the first director of the FBI, "saw the FBI's role as revealing the unseen criminality hidden behind *apparently ordinary* American men and women."[6] Similarly, the social historian David Cunningham has noted that "during his pathbreaking forty-eight year tenure as Director" of the FBI, Hoover "engineered an FBI equipped to engage in a variety of information-gathering techniques, including physical and electronic surveillance, surreptitious examination of mail and trash, burglaries, interviews, and informant infiltration."[7] To scour the trash, to scrutinize the subtleties of sound, movement, and human interaction, to discover the hidden behind the apparently ordinary—to discover the hidden within what avant-gardists frequently called "found environments"—was as much the art of intelligence as it was the foundation of what Kirby called "the New Theatre."

Back in 1965, when Kirby equated that new theater with the improvisational structures of "nonmatrixed performance," he noted that while such structures were typical of the happenings, they did not actually originate with artists like Kaprow, Dine, Grooms, or Oldenberg. They were in fact traceable at least as far back as the Dadaists in Zurich and Paris. There was, of course, much improvisation at the legendary Cabaret Voltaire in Zurich at the outbreak of World War I, but it was hardly confined to the stage. There among the original Dadaist were figures like Emmy Hennings, who apparently was as skilled at forging identity papers as she was at performing cabaret.

Those forged papers gave rough script to the improvised identities that provided safe passage to draft dodgers, pacifists, artists, and political outcasts from Germany into the neutral territory of Switzerland. In a very literal sense, then, the early European intellectual, cultural, and political vanguards had a vested stake in performed, covert identities. Indeed, their very survival amid the madness of World War I often depended on those assumed identities.

It was not only through covert identities that the European vanguards cited by Kirby came to the streets. Within a few short years, "The inclination toward the 'found environment,'" Kirby argued, "was given [further] impetus . . . when Andre Breton suggested [in April 1921] that the Dadaists meet the public in the streets rather than in the theatres and exhibition halls."[8] Rather than waiting for the public to come to the Dadaists through the established conventions of theater, Breton implicitly advocated a mode of performance that would catch the public with its guard down. Inasmuch as Breton was proposing a radically new venue that would make art and performance streetwise and place them where they previously had not been found—a venue where art and life were certain to blur—he might have been surprised to discover that the artful practitioners of undercover surveillance had beat him to the mark. Indeed, they had even beat Zurich's hidden cultural and political outcasts to the mark.

There is a faint echo in the genealogy that Kirby constructed for what he called the new theater to the early history of the emerging practices of undercover surveillance. If Breton's suggestion "that the Dadaists meet the public in the streets" was a call to arms, it was a call not unlike that articulated a century earlier by François Vidocq, the visionary police operative who not only had a short-lived career in the theater[9] but who was also "a master of disguise and an effective dramatic publicist" whose "innovative [undercover] tactics led him to the heart of the burgeoning Paris underworld," where "previous police inspectors had not dared to enter."[10] This was a new theater of operations for police work. As Gary Marx has noted, Vidocq argued that "crime can only be fought by criminals" or by those who could traffic with them undetected. Seeking "to create fear in the underworld that at any moment he might suddenly appear," Vidocq thus "made extensive use of infiltration and deception"[11]—forging a kind of risk-laden theatrical practice among the police that has all of the hallmarks of what Richard Schechner in other contexts has called "dark play."[12] In short, those who would redefine the conventions of theater by taking it to the Paris streets in 1921 were, in no uncertain terms, walking a beat that was already populated with operatives from the police's established new theater of operations. In many respects, these crowded streets foreshadowed the complex modes of performance that would radically redefine theatrical practice in the postwar era as a new or second wave of Western avant-garde performance practitioners united with the political activists of the 1960s and took to the streets in sustained, loosely organized, and creatively improvised demonstrations on behalf of civil rights and in opposition to the Vietnam War.

Political Activism, Police Operatives, and the Radical Redefinition of Theater

As is the case with any constructed genealogy, there is a bit of creative weaving in the revised prehistory to Kirby's notion of the new theater that I am proposing by intertwining the histories of Dada with those of the early forms of surveillance associated with undercover police work. But the justification for weaving these different historical trajectories together has a lot to do with the inopportune timing of the publication of Kirby's widely influential book on the happenings in 1965, which coincided with the *Drama Review*'s publication of its special issue on the happenings in that same year. It was in that issue that Kirby's article "The New Theatre" also appeared. All of these publications—which were in preparation in 1964, and probably earlier—predated the massive escalation of U.S. involvement in Vietnam that began in 1965 and continued through 1969—an escalation that was met by growing opposition and protests, which had a substantial impact on artistic expression as the decade progressed. In this respect, neither Kirby's book nor the essays in the *Drama Review*'s special issue on happenings foresaw the ways in which the artistic developments cited by Kaprow in 1959 and reiterated by Kirby in 1965 would blend with the innovative political activism that took shape in the latter half of the 1960s as the United States committed more and more troops to Vietnam and the country descended into social unrest.

If there was a radical redefinition of theater in the 1960s—if there was a "new theater" to be found—it was arguably a product of this unique mix of radical artistic expression and radical political activism. At one level, one might legitimately argue that this mix was merely a further evolution of the modes of performance that Kirby identified in his essay on the new theater. But there were other sociopolitical forces in this mix as well, and this brings us back to the interwoven historical trajectories of the emerging European avant-garde and the nascent surveillance techniques of undercover police operatives. For just as experimental artists were drawn into the ranks of political activists as the debacle of the Vietnam War unfolded, so too were an astonishing number of undercover operatives and informants working for the FBI. The justification for interweaving the histories of the early western avant-gardes with those of the emerging forms of surveillance associated with undercover police work is thus that the two converged—intertwined as it were—within the spaces of the innovative political activism of the mid and late 1960s where new forms of theatrical and performance expression were emerging. While this convergence suggests the need to rethink our understanding of the radical redefinition of theater in the 1960s, an equally if not more important issue is the way in which this convergence offers key insights into the performative dimensions of surveillance as a sociopolitical phenomenon—insights that resonate well beyond the particulars of undercover surveillance operations. Crucial in this regard is a recognition that by the 1950s undercover surveillance in the United States had expanded beyond Vidocq's earlier focus on the criminal underworld and set its sights on political activists as well.

Undercover surveillance was merely one of the techniques of intelligence, counterintelligence, and disruption used by FBI agents in the 1960s not only against those involved in criminal enterprises but also against individuals or groups who, without having done anything illegal, were nonetheless labeled "dangerous radicals," "subversives," dissidents, or activists and who were thus subjected to surveillance and control. The sheer scope of the FBI's surveillance of presumed radicals and subversives is still difficult to fathom, and, within its own historical context, it matches the level of paranoia that has characterized the post-9/11 era. That scope only began to come to light in the mid-1970s with the formation in 1975 of the Special Select Committee to Investigate Intelligence Activities. Following in the aftermath of Watergate, this committee, better known as the Church Committee, after its chairman, Frank Church,[13] stunned the American public by revealing a long history of FBI abuses, which included among other things revelations that "the FBI had created files on more than one million Americans and carried out more than 500,000 investigations of [so-called] 'subversives' during 1960–74, without a single court conviction"; that "FBI agents had conducted a campaign to incite violence among African Americans"; and that "the bureau had carried out smear campaigns against individuals and groups across the country, simply because they had expressed opposition to the war in Vietnam" or "criticized the slow pace of the civil rights movement."[14] Today, Americans recoil at recent revelations that the NSA monitors the mobile phone calls of countless citizens, but this is nothing new. Among its other revelations, the Church Committee also reported that NSA computers "had monitored every cable sent to or received from overseas by Americans from 1947–75."[15] Granted, this segue into the surveillance tactics of the FBI might seem to be a rather odd departure from Kaprow's sense of the alchemies that would come to dominate the 1960s. But if the radical artistic developments cited by Kaprow tended to blur with the radical political developments that took shape as the 1960s descended into social unrest, so too did the surveillance operations of the FBI deliberately blur into those political developments as well.

COINTELPRO and Its Theater of Operations

This blurring was especially evident in surveillance operations like the infamous COINTELPRO that involved the use of FBI agents as undercover operatives whose assignment was to assume false identities and then infiltrate activist organizations of the New Left like the Students for a Democratic Society (SDS). The official purpose in these operations was not to keep tabs or gather intelligence. Without any sense of due process, the official directives simply assumed that the activists were working against democratic principles. So rather than gathering intelligence, the operations sought to "expose, disrupt, misdirect, discredit, or otherwise neutralize" the effectiveness of New Left activism.[16] At some level, those official goals transcend their own historical specificity and provide us—even now in the twenty-first century—with a rough, unmasked description of the political goals of almost all forms of surveillance more generally. But perhaps the more immediate point is that this infiltration in its use of undercover FBI agents who constructed false identities and passed themselves off as genuine activists was fundamentally intertwined with the period's radical redefinition of theater. Not only were their constructed undercover identities a mode of performance in their own right, but the assignments that the agents of COINTELPRO took on placed them in an influential position within political movements that consciously embraced a keen sense of staged events—indeed, of performance itself—as a crucial component of viable political activism.

At a practical level, COINTELPRO's intense combative engagement with the New Left is an important reminder that even the most celebrated moments of 1960s activism—like the antiwar demonstrations outside of the 1968 Democratic Convention in Chicago—were never fixed within the clear binaries of opposed progressive and reactionary forces that popular historical myths about the period might suggest was the case. As Pnina Lahav has noted in her study of the theatricality of the Chicago conspiracy trial, for example, the case against the Chicago Eight after the 1968 Democratic Convention was presented almost entirely "through the testimony of undercover police agents who had infiltrated the movement's ranks,"[17] and had been a part of the events from beginning to end. Even before the demonstrations at the Democratic Convention in Chicago, those agents were so much a part of movement's activities that it was difficult to tell whether agents like William Frapolly were "employed by the government . . . to spy on the dissenters . . . [or] to encourage them to engage in illegal activity." Indeed, Lahav singles out Frapolly's testimony at the trial. In it, he specifically admits that he enrolled in and then was expelled from Northeastern Illinois State College because his assignment was that of an undercover agent provocateur. In his testimony, Frapolly concedes that he was expelled because he was among a group of students who, in a moment reminiscent of the antics of the Parisian Dadaists forty years earlier, stormed the college's stage while the college president was speaking and in a grand provocative gesture literally picked him up and threw him from the stage. Under cross-examination from the defense attorneys for the Chicago Seven, Frapolly admitted that, while working undercover, he even participated in the discussions that led to the assault.[18] Just how extensive his involvement with the planning of the assault was remained a matter of dispute even in court, but he clearly had a role—indeed, played different simultaneous parts—in the events that transpired on the stage at the college and in the streets of Chicago outside of the Democratic Convention in 1968.

Frapolly's work as an agent provocateur was by no means the sole instance of the blurring of the lines between left and right, between right and wrong, or, I would add, between art and life during the antiwar activism that came to a head as the decade progressed. Writing specifically about Chicago's undercover operation from 1966 to 1976, the political historian David Garrow remarks, for example, that in Chicago alone, the FBI "operated at least 5,145 security, radical or 'extremist' informants, and paid those sources a total of at least $2.5 million dollars over the course of that decade." He argues furthermore that "when extrapolated nationally" the number of operatives who were listed by the Chicago FBI—and this only includes those about whom records were kept—not only suggests "an informant corps of perhaps some 100,000 individuals over the course of those ten years," but also "budgetary expenses for them that reached easily into the tens of millions of dollars."[19] Within their historical context, these kinds of numbers are rather staggering, particularly since they only represent the exposed edge of a operational force that by definition the FBI sought to conceal. But the existence of such operations ought not come as a surprise. As David Cunningham has noted, political activism, first and foremost, is always "an interactive struggle between challengers and authorities." The results of activism are thus "highly conditioned by the actions of both sides of the equation."[20] While Frapolly's testimony certainly attests to this very point, in many respects it is of course rather obvious that the results of political engagement are as much a product of "protest actions" as they are of the "authorities' facilitation or repression of these actions."[21] Obvious though this may be, however, there has been very little discussion of it among theater historians who have counted the radical political activism of 1960s as a crucial component of an emergent American postwar avant-garde and of the period's radical redefinition of theater more generally.

Working with a broad understanding of the fluid spaces of theatrical practice in the 1960s, theater historians and performance studies scholars readily acknowledge the theatricality—indeed the fundamental redefinition of theater itself—that was to be found in moments like the Yippies' orchestrated attempt to levitate the Pentagon in 1967, the protests engulfing the 1968 Democratic Convention, the Grosvenor Square demonstration in London of March 1968, and not least of all the massive demonstrations in Paris in May of that same year. Though writing specifically about the May 1968 demonstrations in Paris, Jean-Jacques Lebel, for example, closed the decade in 1969 much in the way that Allan Kaprow had opened it in 1958: by proclaiming that "the old avant-gardist dream of turning 'life' into 'art,' into a collective creative experience, finally came true."[22] By the end of the decade, the "unheard-of happenings and events" that Kaprow had predicted would be "seen in store windows and on the streets" had morphed into the broken glass and the cobblestone barricades of what one might legitimately call the production, appropriation, and creative use of detritus and junk in radical political warfare. In fact, one of the most notorious acts of creative activism was literally that: late in 1968 during a sanitation workers strike in New York City, the group calling themselves the Up Against the Wall Motherfuckers collected a truckload of the trash that had accumulated on the city streets during the strike, "dumped it on the steps of Lincoln Center"—which at the time was hosting a gala event—and then distributed flyers announcing that they were proposing "a cultural exchange": "garbage for garbage."[23] While such acts were described in the press as "Garbage Guerrilla," they just as easily fit the mold of what Lebel, in an echo of the San Francisco Mime Troupe's R. G. Davis, simply called "Guerilla Theatre." So too were they part of a tradition that the Motherfuckers traced all the way back to Dada.[24] All of this contributed to a political stance espousing the notion that the only viable theater was theater in the streets. Ironically, this was an opinion shared by the undercover operatives who walked arm and arm with those whom they sought to "misdirect" and "discredit" through clandestine modes of performance that smudged the blurring of art and life with calculated deception and betrayal not only of the activists whose colleagues they pretended to be, but also of the democratic principles in whose name the agents of COINTELPRO ostensibly had been deployed.

Back in the late 1990s, critics like Andrew Hunt noted how few writers and critics had "acknowledged the devastating impact of the

FBI's Counter-Intelligence Programs . . . on political activism in general and the New Left in particular."[25] But there is need to go even further. Indeed, if one were looking for a plausible way to establish the parameters of the 1960s as a long decade and as a decade devoted to a radical redefinition of theater, a good place to start would be by suggesting that the performative 1960s began with the creation of COINTELPRO in 1956 and ended with its dismantlement after it was exposed in 1971. The plausibility of this suggestion is only strengthened by the fact that COINTELPRO's beginnings in 1956 meant that its undercover operations—its clandestine improvisational performances—were not really a response to the New Left or to the New Left's creative embrace of performance and theatricality as strategies of activism. On the contrary, through COINTELPRO the FBI had cultivated a mode of improvisational, experimental performance that in a very literal sense was "in advance of"—that was avant-garde to—the New Left's radical redefinition of theater in the streets. In this respect, David Cunningham was more correct than he perhaps realized when, in the introduction to his book *Something's Happening Here*, he claimed that COINTELPRO cast "the key players and events in what was perhaps the most sweeping and contentious domestic political drama of the past century."[26]

Keeping in mind that Frapolly was merely one in a long list of agents whose assignment was to "misdirect," "disrupt," and "discredit" the activism of the New Left, I would like to suggest, first of all, that if there is a compelling case to be made that the political activism of the 1960s was inherently part of the decade's radical redefinition of theater, then the undercover operatives who sought to disrupt that activism were as much a part of that redefinition as the activists themselves. Given the multiple layers of performance in which they were simultaneously involved, there is in fact a plausible, albeit disquieting, case to be made that those agents practiced a substantially more sophisticated and nuanced mode of improvisational and experimental performance than that practiced even by the most skilled performance-oriented activists like Abbie Hoffman and Jerry Rubin.

Hoffman, a founding member of the Yippies and one of the Chicago Eight, is perhaps best remembered for orchestrating mayhem at the New York Stock Exchange on August 24, 1967, when he and a small group of activists threw three hundred single dollar bills onto the trading floor of the Exchange from the balcony of the visitors' gallery, an act that forty years later the journalist James Ledbetter called "perhaps the most striking act of guerilla theatre in American history."[27] Vincent Cannato recalls in his book *The Ungovernable City* that "street theatre was Hoffman's stock and trade" and that "in another incident, . . . Hoffman and his friends claimed to have created a special foam which, when applied to a person's skin, would make that person automatically have sex with the next person, male or female, the 'foamed' person came in contact with."[28] Since the Yippies already had a reputation of experimenting with all sorts of chemical drugs, this claim was given rather surprising credibility, and, as Cannato notes, Hoffman created panic once again when at a local precinct he sprayed a police sergeant, who immediately began "to moan and yell, panicked at the thought that the foam would automatically force him to have sex with the police lieutenant standing nearby."[29]

Such antics were arguably met in kind by COINTELPRO agents who, among a long list of schemes to disrupt the New Left, hatched a plan to infiltrate an antiwar demonstration and distribute among the activists oranges that had been injected "with powerful laxatives."[30] At the level of mere political mischief and antics, there was of course a kind of adolescent quid pro quo in all of this. Humorous though such antics might be however, they pale in comparison to the more subtly calculated strategies of deception and diversion practiced by COINTELPRO agents in their efforts to quell not only the antiwar movement but the civil rights movement more generally. Ironically enough, the political theater of those strategies has a lot to do with the fact that they played as much among the FBI's COINTELPRO program against the Ku Klux Klan as they did against civil rights activists per se.

FBI efforts to discredit major civil rights figures like Dr. Martin Luther King Jr. are well known. Yet those efforts coincided with a COINTELPRO program that not only targeted the Klan but that actually predated COINTELPRO infiltration of New Left antiwar activism. As Cunningham has noted, all of these programs first came to light in March 1971 when a group of antiwar activists who called themselves the Citizens' Commission to Investigate the FBI broke into FBI offices in Media, Pennsylvania, stole files on the FBI's undercover operations against U.S. citizens, and in an act that could easily have served as inspiration for Edward Snowden forty years later "gradually leaked them to various media outlets in succeeding weeks."[31] While this act of civil disobedience ultimately led to the collapse of all COINTELPRO programs, one of the enduring puzzles of the information that the activists leaked concerned the implications of the substantial infiltration of FBI agents into various "klaverns" of the Klan. At one level, infiltration of the Klan ultimately implicated the FBI in events like the infamous brutal assault on the Freedom Riders as their bus rolled into Birmingham in 1963, and perhaps even more disturbing, one of the Klansmen involved in the murder of civil rights activist Viola Liuzzo in March 1965 was an FBI undercover operative by the name of Gary Thomas Rowe. In his book *The Informant*, political historian Gary May provides a careful reading of Justice Department records and presents a compelling argument, for example, that in its efforts to protect Rowe's cover, the FBI became accessory to some of the most egregious crimes of the civil rights era.[32] But at another level, these two events were anomalies in the political theater that Hoover orchestrated through the COINTELPRO program against the Klan. If Abbie Hoffman was a master of street theater, Hoover was no less a master of political theater and intrigue.

What we also know is that Hoover was no wallflower when it came to racism. His own pronounced bigotry and his personal hatred of Martin Luther King Jr. are part of the ugliest and enduring marks of his legacy. So it is no small matter that between 1964 and 1971 the Bureau was running simultaneous COINTELPRO programs against the civil rights movement and "Klan-related groups seeking to maintain the existing racial order in the South."[33] But if nothing else, this seeming contradiction ought to cause enough pause to give us opportunity to examine it for the brazen form of theatricality that it was. Not only was there extensive use of undercover infiltration into the Klan by agents like Gary Thomas Rowe, but those operatives were being played in a theater of operations that was much more complex than the stage that they could see. Scholars like William Keller have argued that COINTELPRO programs against the Klan were set in motion by a staunchly reactionary and conservative Hoover in a shrewd ploy to curry favor with a liberal administration and thus "to gain the degree of insularity and autonomy" that he wanted for its other programs.[34] Kenneth O'Reilly has noted furthermore that

"Hoover saw the Ku Klux Klan as another subversive threat to the peace and stability of middle America" and believed, in a grotesque moment of exposed racism, that the Klan was giving a bad name to "the anti-civil rights movement."[35] Perhaps most important of all, because of "the liberal support [that they] received for COINTELPRO–White Hate Groups, Hoover and the FBI achieved sufficient insularity and autonomy to be able to establish subsequent counterintelligence programs against domestic targets without the approval of Congress or other actors outside of the FBI."[36] In short, the COINTELPRO program against white hate groups substantially widened the range of groups that Hoover was subsequently able to define as "subversive" and thus target. By going after the Klan, Hoover gained leave to conduct operations against fully legitimate and legal political groups on the left with whose politics he disagreed.

The multiple layers and players of political theater in these strategies ought not be overlooked. But the point here is not merely that contemporary scholars writing on the performance legacy of the 1960s today need to recalibrate their understanding of the dynamics of the period's radical redefinition of theater in ways that take into account the performative modes of surveillance and counterintelligence that the state successfully wove into political activities of the New Left, of the civil rights movement, or even of the Klan. We would do well also to consider the ways in which that legacy of infiltrated activism has evolved over the past half century and of the ways that this evolution might provide us insight into the performative dimensions of surveillance more generally. There is much to be gained by this latter consideration. Simple though the use of undercover operatives may be compared to the advanced surveillance technologies available today, that simplicity contains key precepts that continue to influence and structure surveillance as a social-political phenomenon.

Surveillance, Performance, and Directing

Among those key precepts are elements of performance that extend well beyond the use of fabricated identities. In this respect, I disagree with Gary Marx's argument in what is one of the truly generative books for surveillance studies, his 1988 book *Undercover*. There, he subtly acknowledges the performative dimensions of undercover surveillance but mistakenly cordons them off from other forms of surveillance, or at least from the forms of surveillance that were already emerging at the time his book was published. Nowhere is this more evident than in the privileged sense of influence he ascribes to undercover operations compared with more technologically based forms of surveillance. Arguing, for example, that "the dangers from intrusion are much greater with undercover . . . [operations] than with electronic surveillance," Marx maintains that "the actions of a decoy and the creation of a tempting opportunity can affect conversation and behavior in ways that a hidden recording device never can."[37] It is worth dwelling momentarily on the effects that Marx cites.

Such effects are performative because actions taken to create "tempting" opportunities or to "affect conversation and behavior" orchestrate—indeed, they direct—human behavior and interaction. The question is whether, in the larger sphere of surveillance activities, such effects are unique to undercover operations. Maybe this was true in the late 1980s, but it is certainly not the case now. A powerful illustration of this very point can be found in the November 2013 PEN America Center report entitled *Chilling Effects: NSA Surveillance Drives U.S. Writers to Self-Censor*. The PEN America Center "is the largest branch of PEN International, the world's leading literary and human rights organization." As is well known, "PEN works in more than 100 countries to protect free expression and to defend writers and journalists who are imprisoned, threatened, persecuted, or attacked in the course of their profession." In its November 2013 report, the PEN America Center assesses the impact that NSA surveillance is having on American letters and journalism. It notes that American writers "now assume that their communications are monitored" and that as a consequence writers have literally modified their conversations, correspondences, and behavior. The report notes that research indicates American writers are increasingly reluctant "to write or speak" or even "to pursue research about . . . subjects" that, because of their politically controversial nature, might trigger what we might describe as COINTELPRO-style government actions against the writers themselves, their families, or their colleagues and friends. Indeed, the report notes in particular that American writers have become progressively more reluctant "to communicate with sources, or with friends abroad, for fear that they will endanger their counterparts by doing so."[38] It is hardly a stretch to assume that such tendencies among writers can also be found among artists, activists, and citizens in general.

Taking broad exception with Marx's argument, then, I contend that while the use of undercover operatives is arguably the most rudimentary and oldest form of surveillance used today, its performative foundations actually extend to almost all forms of surveillance, regardless of whether one speaks of the increasingly prevalent presence of CCTV cameras, the widespread collection of data and the construction of data doubles, or the covert and overt tracking of cell phones, email, and social media. Each of these forms of surveillance seeks to orchestrate and affect human interaction. Each seeks to direct our behavior. At some level, we are generally conscious of all of them, and they thus influence us, regardless of whether the mode of surveillance is "hidden" or visible. Like the operations of COINTELPRO, the multiple forms that surveillance techniques and technologies have today are all designed to influence and direct us—even if that influence or direction might take the form of what PEN America calls a "chilling effect." In this respect, the performative dimensions of undercover operations not only point toward a common underlying structure to surveillance more generally; they also point toward something that amounts to a radical shift in our understanding of the principle operative structures of surveillance as such.

It's not merely that the strategies of disruption, misdirection, manipulation, and subtle coercion that the agents of COINTELPRO used did not actually end with the termination of COINTELPRO in 1971, or even that those strategies are operative in almost all forms of surveillance today. It's that the very term "surveillance" itself is a bit of a misnomer that distracts from the performative realities of surveillance as a sociopolitical phenomenon. Those realities run counter to the term's suggestion of passive observation, of unobtrusive intelligence collection, or, more simply put, of "watching." Despite what one might expect when thinking about surveillance's performative foundations, surveillance is not really aligned with the principles of spectatorship. Surveillance is primarily a mode of directing. Its objectives are focused less on observation and intelligence gathering than on influence, persuasion, control, containment, and coercion. Its goals are to mediate and thereby shape the parameters of human events, interaction, communication, and exchange.

Inasmuch as the objectives and goals of surveillance are focused on directing, the crucial critical questions about surveillance arguably are not really covered in the well-known Latin formulation: *Quis custodiet ipsos custodes?* or its English variant, "Who watches the watchers?" Nor can this formulation simply be modified and distilled into the parallel interrogative: *Quisnam directus presul?* or "Who directs the directors?" These are important questions, of course. But equally important are queries about whom the directors direct, and about how they direct. Back in the late 1980s, when Gary Marx first published *Undercover*, he touched upon this very issue in the chapter devoted to the "unintended consequences" of surveillance on its targets. There, in an implicit reference to directing, he warned how easily "investigation becomes instigation" and how, when targets encounter particularly aggressive undercover operatives, criminal actions "may result from intimidation rather than free choice." Trickery, which is "inherent in the tactic" of undercover surveillance, is no less potentially problematic than intimidation itself, Marx argued, since "both elicit behavior" that for all intents and purposes is "nonvoluntary."[39] Both, I would add, are also performance strategies and techniques that fall within the parameters of directing.

The Informant as Director: The Case of Brandon Darby

It is this sense of directing—and of performance that has blurred into a coercive and dubious kind of provocation—that, long after the official end of COINTELPRO, continues to characterize undercover surveillance operations against political activists in the United States and Europe. Indeed, the notion that surveillance is primarily a mode of directing offers a critical insight into the fundamental issues haunting one of the most controversial undercover operations against U.S. activists in the last decade. This operation, which coincided with a forceful police crackdown on activists who came to St. Paul, Minnesota, to protest the 2008 Republican National Convention, involved a highly respected radical political activist by the name of Brandon Michael Darby. Darby had come to prominence as a progressive activist, first in Austin, Texas, and then in New Orleans. There he was among the founders of Common Ground Relief, "a community effort focused on restoring neighborhoods, defending residents' rights, and trying to rebuild the community" in the aftermath of Hurricane Katrina. As Diana Welch reported in a 2009 article for the *Austin Chronicle*: "Darby was featured in national interviews about post-hurricane New Orleans and more particularly had a reputation for defying authority and especially cops."[40] All of this seemed to make him an unlikely candidate for the police informant that he turned out to be. But like the term "surveillance," "informant" here too is a bit of a misnomer. "Agent provocateur" comes to mind as viable alternative, and given Darby's penchant for assuming leadership roles, so too does the term "director."

During his time at Common Ground Relief, Darby ultimately became divisive enough to drive "out 30 volunteer coordinators."[41] This kind of divisiveness, which has all too familiar echoes of COINTELPRO's strategies of disruption and misdirection, has led activists like the former Black Panther Party member Malik Rahim to conclude that Darby was working for the FBI "for at least part of the time he worked for Common Ground."[42] Darby had befriended Rahim, a founding member of Common Ground Relief, who was well known and respected in activist circles in New Orleans. Initially, they worked well together. But Darby gradually became disruptive, driving a wedge, for example, between Rahim and one of his closest allies, the activist Lisa Fithian. As Rahim notes in retrospect, Darby "was doing everything you're supposed to do as a government agent in that situation. Divide and conquer."[43] Indeed, FBI files subsequently obtained by Fithian indicate that during this period she was one of Darby's primary targets, and he was reporting regularly on her to the FBI.[44] Other FBI files on Darby's former close friend, the Austin-based activist Scott Crow (also a founding member of Common Ground Relief), indicate that Darby was working for the FBI at least as early as 2006—at the same time that in two separately corroborated accounts he reportedly attempted to convince Crow and fellow Austin activist Karly Dixon "to help him burn down an Austin bookstore affiliated with right-wing radio host Alex Jones." "The guy was trying," Crow claims, "to put me in prison."[45] As was the case with Fithian, Darby was also reporting to the FBI about Crow. As Diana Welch has noted: "During 2007, Darby secretly informed on Crow's whereabouts and actions."[46] In this respect, Darby was apparently part of an orchestrated effort. In an echo of what Allan

Kaprow had called the alchemies of the 1960s, Crow's FBI files also indicate that among its multiple surveillance tactics against Crow the FBI liked to rifle through his trash.[47] "COINTELPRO did not end in the 1970s," Crow argues, "it simply diversified and changed names."[48] To Crow's observation, I would only add that alchemy by any other name is still alchemy.

While Darby is dismissive of claims that he was trying to entice Crow into criminal behavior,[49] he readily admits that he played an instrumental role in the criminal case against two young Texas activists, David McKay and Brad Crowder, who both ultimately pled guilty and did time in prison on charges that they constructed Molotov cocktails in St. Paul during the 2008 Republican National Convention. No one contests the fact that McKay and Crowder built the Molotov cocktails. The real questions are how they came to do so and whether Darby encouraged them to build and use firebombs. Indeed, these questions are central to the controversy surrounding Darby's precise role as a paid undercover operative in the events leading to their arrest. At the very least, it was a two-faced role, and in this respect, the controversy regarding Darby is positioned on the lines that ostensibly separate investigation from instigation; that supposedly discriminate an informant from a provocateur; and that hypothetically distinguish the notion of surveillance as a mode of passive spectatorship from that of surveillance as a mode of active directing. All of these, I would suggest, are false distinctions since in each instance discretely doing the latter is greatly facilitated by hyping up the former. But more than anything else, I would suggest that the controversy regarding Darby and his ethically suspect duplicity in the case against McKay and Crowder is a snapshot of the Janus face of surveillance as such. It is never neutral. It is always a profoundly disruptive and influential force, shaping the events it disingenuously purports only to observe, surveil, and record.

Darby's specific involvement with Crowder and McKay began in the months leading up to the Republican National Convention when he was asked by the FBI to attend a meeting at the left-leaning Monkey Wrench Bookstore in Austin, Texas. That meeting was one among a number of organizational meetings sponsored by radical activists in Texas and across the country who were intent upon disrupting the Republican convention, and Darby's presence at this particular meeting could not have gone unnoticed. His outspoken reputation accompanied him. In Austin's more radical circles, Darby was a well-known, street-wise, and charismatic figure: a seasoned activist who had a reputation for not shying away from confrontational politics or acts of civil disobedience. Lisa Fithian recalls, for example, that Darby was never "squeamish about violence. He owned guns and cultivated his reputation as a hotheaded, militant revolutionary." She also notes that at the time of David McKay's trial "at least a half a dozen people were prepared to testify, under oath and with some risk" to themselves and others, that Darby "had approached them with proposals to commit robbery or arson."[50] In fact, following the first trial of David McKay, which ended in a hung jury, federal prosecutors filed a motion to keep defense attorneys from telling jurors in the retrial that Darby had "advocated using arson to fight developers who were forcing low-income residents from their neighborhoods," which under oath Darby admitted during cross-examination in the first trial.[51]

When he showed up at the Monkey Wrench Bookstore early in 2008, Darby was already in his thirties, a veteran activist with a solid decade of street demonstrations behind him. He was ten years older than McKay and Crowder, who were gearing up for their first major demonstration. In break-out sessions, Darby was paired with McKay and Crowder and another activist by the name of James Clark, who like McKay and Crowder was in his early twenties. Clark knew Darby and his reputation and recalls that Darby dominated the conversation. By Clark's account, Darby was certainly playing a preemptive role that channeled the conversation in the directions he wanted. Reporting back to his FBI handlers, Darby wrote in an email that he had told the others that he "wasn't interested in being a part of a group if . . . [they] were going to sit and talk too much." Referring specifically to the Republican National Convention, Darby wrote in that same email: "I stated that I was gonna shut that fucker down."[52] These kinds of statements and the reputation that Darby brought to the table don't square too well with his subsequent claims in numerous interviews that his "role" as an informer was "to provide information [to the FBI], and try to not influence—not to be more radical but not to be less radical. To fit in."[53]

Given his reputation for having a very assertive and strong personality, Darby could only fit in by being the person that his reputation led others to expect: only by being an outspoken and commanding figure, a voice of experience. Writing later from prison, David McKay told Pamela Colloff in a piece published by the *Texas Monthly*: "Many questions about [Darby's] motivations to crucify us go unanswered to me. For he was . . . someone we followed and looked up to, who proclaimed himself a person of guidance and a person of action."[54] This is the role that Clark, McKay, Crowder, and others all agree he assumed. It is a role that gave Darby a substantial amount of influence, particularly over individuals like McKay and Crowder, who were inexperienced and susceptible to the counsel offered and subtle direction provided by a more experienced and respected activist—the kind of direction that Darby arguably provided by implication and example if not by actual word. If, as Darby claims, his job was to be just as radical as everybody else in the room, the example he set as an experienced activist was enough to have a profound influence on novice activists learning the behavioral parameters of radical civil disobedience. As a result of his status, reputation, and the sheer force of his personality, Darby was not—indeed, could never simply be—neutral. He was a disruptive and influential force, shaping the events he disingenuously claims only to have observed and reported. He was, in short, an example of surveillance personified: an influential and directing force.

Those familiar with the case of McKay and Crowder know that in preparation for the street confrontations with police deployed around the Republican convention in Minneapolis / St. Paul, they had constructed body shields out of split plastic street drums (i.e., barrels) in order to protect themselves against police batons. They transported the shields to St. Paul in a U-Haul but never had a chance to use them. Working in response to a tip given to them by Darby—a tip that had a direct and influential effect on the events that followed—the St. Paul police made a decision of questionable legality: they broke into the parked U-Haul trailer and confiscated the shields while McKay, Crowder, and the others were meeting with fellow activists. Frustrated by this turn of events and with the very heavy-handed tactics of St. Paul's riot police on the streets, McKay and Crowder then purchased the necessary materials and constructed Molotov cocktails in the house where they were staying. Although Darby apparently did not specifically tell them to construct the firebombs, there is considerable debate about the extent to which he strongly encouraged and cultivated such behavior. Austin-based activist Gabby Hicks, who was among the group in St. Paul that included Darby, McKay, and Crowder, has argued, for example: "As an older seasoned activist, Darby had a lot of sway over Crowder and McKay, making them susceptible to his often militant rhetoric," and she maintains that Darby "was always the one to suggest violence when the rest of us clearly disagreed with those strategies."[55] Moreover, there is considerable evidence in the text messages that Darby sent to McKay once he had learned of the Molotov cocktails' existence that he tried to set McKay in motion to use them. Indeed, McKay, who like Crowder is now out of prison, confronted Darby in a public debate in September 2013 about this very issue. After Darby claimed that it isn't fair to say that he encouraged McKay, McKay responded: "You came to me, after I had been arrested [i.e., detained with other demonstrators for civil disobedience], after the protest, after I was emotionally exhausted after my disappointment because of things I've experienced and saw from a lot of things. . . . You were trying to antagonize me to go use them."[56]

That antagonism came after other members of their own group like Gabby Hicks had confronted Crowder and McKay and after they had decided not to use the firebombs. Indeed, it is important to see Darby's claim that his role was merely "to provide information, and try to not influence—not to be more radical but not to be less radical" in the larger context of how he learned about the Molotov cocktails in the first place. In Jamie Metzer's documentary film *Informant* (2013), Darby recalls that he learned of the Molotov cocktails when he received a text message (probably from Hicks) stating that the group was "in a big fight" because McKay and Crowder had purchased the material to construct Molotovs. Darby says in the film: "I really wanted to just grab . . . [McKay and Crowder] and be like, 'Hey, what you're doing is so stupid. You're going down the wrong path. This is ridiculous.'" Only he couldn't, he claims, because he had to "try to not influence," because he had to "fit in," and because it was not his "place to talk people out of things."[57] And yet, all of this presumes a kind of stability and uniformity of opinion among the activists that is simply a convenient and disingenuous piece of fiction. Darby only learned of the firebombs because the group was challenging—and in fact was successfully challenging—Crowder and McKay to reconsider what they were thinking of doing. To be no more or no less "radical," to "fit in" would have meant in this instance to choose a side in the argument that had erupted in the house where McKay and Crowder were staying. The majority opinion in that argument, the "measured" radical opinion, was against the use of the Molotovs. Darby chose not to weigh in with the majority. That was a significant and influential choice since it suggested that, as a seasoned activist, he was of a different opinion.

When all was said and done, the FBI could not find evidence to charge McKay or Crowder with conspiracy or with intent to actually use the Molotov cocktails.[58] The two were only charged with possession of firearms that were not registered to them (the Molotovs technically qualified as firearms). Although the arrests were played up in the media as a foiled bomb plot, the actual case against McKay and Crowder didn't live up to the hype. It was a poor return shrouded in theatrics: a kind of show trial that deflected attention from two major issues. First, McKay and Crowder arguably ended up in prison not out of a larger sense of justice for their crime but out of a need for the FBI to justify the substantial amount of resources devoted to supporting Darby's undercover work—resources that were allocated long before the Republican National Convention, that produced an operation large enough to cover activities in Louisiana, Texas, and Minnesota, and that targeted individuals "who were engaged in lawful activism" and who had "no plans to attend the Republican Convention."[59] Given its scope, it was an operation that—despite Darby's subsequent grandstanding for "tea party" activist gatherings and other conservative audiences—had embarrassingly little to show for itself except for its own not so neatly disguised political agenda. Second, Darby has described McKay and Crowder as "collateral damage,"[60] a description that is worthy of note because it subtly reminds us that they were never specific targets of an undercover operation against criminals. They were simply small and insignificant players who were coincidentally caught in the net of a larger operation targeting a wide range of activists more generally—activists that included the likes of Rahim, Fithian, Crow, Dixon, Clark, Hicks, and others—all of whom were subjected to surveillance and many of whom, it would appear, were unsuccessfully being directed toward criminal activity by the FBI. This was not because they were involved in or inclined toward criminal enterprises but because of their political views and activism: because, to borrow a term straight from Hoover's playbook, they were considered "subversives." In this regard, the expansive operation leading to the case against McKay and Crowder carries a very clear echo of COINTELPRO's concerted efforts to "expose, disrupt, misdirect, discredit, or otherwise neutralize" the effectiveness of New Left activism.[61]

Undercover Domesticity: Surveillance, Privacy, and Playing House

Both COINTELPRO and the FBI operation behind Brandon Darby involved a political targeting of citizens for surveillance rather than a targeting of crime. Such strategies consistently place surveillance at odds with issues of free speech, freedom of assembly, and, above all, with basic privacy. There is ample evidence to suggest that the profound intrusion of COINTELPRO agents into the private lives of ordinary citizens set a precedent that, as the case of Brandon Darby illustrates, not only continues to reverberate today but is arguably also one of the primary models after which virtually all forms of surveillance techniques and technologies are fashioned. It is a model grounded in a complex mode of performance with wide-ranging political implications. With regard to issues of political diversity in a democratic society, the journalist Will Potter has rightly argued, for example, that "by gathering information on law abiding activists and then defending his actions as stopping violence, Darby contributes to the public perception that political dissent is criminal, which has a chilling effect on free speech."[62] But the larger issue here, I would suggest, is that Darby is merely a personified example of a characteristic that is true of surveillance more generally. What Potter's comments subtly indicate, particularly in his reference to "public perception," is that as a sociopolitical force, surveillance is also a technique of persuasion, and if we ask how surveillance performs—how and whom it directs—the answer points well beyond activist communities toward the production or manufacturing of consent and conformity among the body politic more generally. The nominal targets of surveillance are thus but a fragment of the performances that surveillance orchestrates and directs.

While Darby performed acts of deception by pretending to be other than he was and while evidence suggests he sought to entice, to tempt, and hence to steer and direct activists toward counterproductive if not criminal acts, the larger surveillance operation of which he was a part implicitly directed public opinion toward a vilification of any action or idea that does not perpetuate a preexisting, normative political and social order. Indeed, Darby's current activities on the conservative lecture circuit and as managing director of Breitbart Texas are arguably an extension of process of calculated vilification. In this respect, the familiar cliché that "only those who have something to hide need to be concerned with surveillance" is a deeply misleading—if not blatantly dishonest—assertion. For it deceptively places surveillance in the sphere of investigation rather than of instigation and in sphere of detection rather than of direction. Mix the two if you want, but surveillance is always already at least as much about instigation and directing as it is about investigation or detection.

Beyond attempts to influence public perception, the techniques and technologies of surveillance—even basic techniques like undercover surveillance—increasingly draw the general public into their operations, and it is this expansiveness that underscores the extent to which the performative foundations of surveillance intersect with fundamental issues of privacy, even at the domestic level. Legal scholars remind us that when it comes to the law, the concept of privacy is notoriously elusive. But it should come as no surprise that the earliest forms of legislation protecting privacy are intertwined with anxieties about surveillance. John Soma and Stephen Rynerson note, for example, that the Code of Hammurabi, which was drafted in the Babylonian Empire almost eighteen hundred years before the birth of Christ and which is "one of the earliest legal texts" to address a principled right to privacy, "made it a crime to break a hole through the wall of another's house."[63] While it is tempting to find the early foundations of theater's fourth wall in the articulation of this crime, the violation described here is one that historically legal scholars have interpreted as an intrusion upon what might legitimately be described as protected spaces of rehearsal.

Noted legal historian and theorist Ruth Gavison argued as much in one of the most important articles on privacy in the last fifty years. In "Privacy and the Limits of the Law," Gavison wrote in 1980 that "privacy is important in those areas in which we want a refuge from pressures to conform, where we seek freedom from inhibition, the freedom to explore, dare, and grope," and she added that "invasions of privacy are hurtful because they expose" and inhibit us in those moments of exploration and because they thus diminish "our capacity to have meaningful relations with others."[64] Among the noteworthy arguments emerging from this line of thought is the equation of the capacity for developing meaningful relationships with the protection of spaces for uninhibited exploration. Such spaces are the spaces of rehearsal, and, at some level, they are arguably akin to the performance laboratories and ensemble workspaces that are among the most enduring legacies of 1960s theater. At the very least, they are the kinds of spaces to which Erving Goffman referred in his now classic study *The Presentation of Self in Everyday Life*—that germinal text of performance studies that uses a theatrical metaphor to understand everyday life and which coincidentally was published in the same year that COINTELPRO was inaugurated. In its preface, Goffman noted: "The stage presents things that are make-believe; presumably life presents things that are real and sometimes not well rehearsed."[65] It is a small step from the realm of the not-so-well-rehearsed to the freedom for uninhibited exploration that, according to Gavison, constitutes the spaces that we commonly associate with privacy and the protected realms of the domestic sphere: that is, the spaces that, in a manner of speaking, are behind the fourth wall through which we are forbidden to bore. There, as in the spaces of rehearsal, trust is essential, and private spaces for trial and error are vital to personal and interpersonal development. So when those realms are intruded upon in acts of surveillance, the sense of violation is understandably strong; the potential for lasting damage is real; and, the justification for intrusion must therefore be ethically, morally, and legally sound.

Not surprisingly, the deeply problematic performative legacy of COINTELPRO is to be found in such intrusions as well. Indeed, that legacy is evident in ways that add a further and in many respects much more disturbing layer of complexity to the notion that surveillance is a mode of directing—particularly if we keep in mind that good directing is as much about establishing and orchestrating context as it is about guiding individual actions. In undercover operations, establishing context often means establishing relationships with individuals who thus unknowingly provide credibility for an operative who is targeting others, and the credibility derived from those relationships frequently has a lot to do with their level of intimacy since intimacy suggests trust and, by extension, trustworthiness. In his article "Under-the-Covers Undercover Investigations: Some Reflections on the State's Use of Sex and Deception in Law Enforcement," Gary Marx recalls, for example, that among COINTELPRO's numerous questionable undertakings was an attempt to infiltrate a particular "student milieu" that FBI agents thought might be affiliated with "the Weather Underground faction of Students for a Democratic Society."[66] As part of this operation, an FBI agent consciously sought out and developed a relationship with a student activist whom he ultimately got pregnant, convinced to get an abortion, and then deserted when the operation proved unsuccessful. The invasion of this young woman's privacy is as hard to dispute as it is to defend, especially when one considers that the operation led to no charges against her or against anyone else in her group. The woman was not even the specific target. Marx also recounts a similar investigation with equally vacuous results in Los Angeles in the 1980s. This latter investigation led to a seventeen-month relationship that was used "to establish . . . [an agent's] credibility" within a "Maoist political organization," and yet when all was said and done, the agent testified that "he never heard anyone in the organization talk about committing crimes nor did he see any weapons."[67] At the most basic level, examples such as these add a degree of nuance to the term "collateral damage" that overshadows even Brandon Darby's rather cynical use of the term to describe the fate of David McKay and Bradley Crowder.

The experiences of the women cited in Marx's examples offer deeply troubling illustrations of how individuals who have nothing to hide and so ought to have nothing to fear, do in fact have to fear not only that they might become the targets of undercover surveillance operations because of their political views but also that they might involuntarily and unknowingly be cast as bit players in an operation in which they are not even a target but for which they may serve as a useful prop. To say then that surveillance is not really about spectatorship but about directing is only to scratch the surface of the expansive reach of its performance stratagems. The legacy of COINTELPRO reminds us that surveillance may be used to direct individuals or groups; it may be used to manufacture public opinion and consent; and so too may it be used in ways that have no justifiable basis for intruding into the lives of individuals whose privacy it violates as a means to an end that has nothing to do with them. This is not just a gross violation of the privacy of those suspected of criminal activity, or of those who for whatever reason have been labeled "subversives," or even of those who have been targeted by private companies for profit. It is a profound illustration of surveillance's expansiveness and of how, in a manner consistent with the radical redefinition of theater in the 1960s, surveillance erases the lines distinguishing participants and bystanders. For this kind of surveillance, all of the world is in fact a stage and everyone is vulnerable to being cast as minor players in a plot and ploy in which they have not sought a stake and in which they shouldn't even have a part.

The justification for such invasions of privacy is difficult to fathom, even when it comes from those who bear responsibility for the operations in the first place. Daryl Gates, who was Los Angeles's chief of police in the 1980s when the above-mentioned operation was undertaken, once defended the use of such deceptive relationships as a necessary strategy for protecting the lives of undercover operatives.[68] Of course, the irony of this defense is that it is wildly at odds with the Law Enforcement Code of Ethics directive "to protect the innocent against deception,"[69] and Gates's defense of deceiving law-abiding citizens and invading the most intimate realms of their privacy thus begins sound a lot like the army major in Vietnam who justified the leveling of the city of Bến Tre by telling the press: "It became necessary to destroy the town to save it."[70] The logic in this particular case is presumably that it is necessary to deceive the public in order to protect it from deception. Beyond the obvious irony, however, lies the very serious risk that, rather than protecting the life of an undercover operative, the constructed relationship might actually jeopardize the unknowing and innocent partner's life if and when the surveillance operation comes to light. Yet even this danger begs the more fundamental question about the assumption that the forces of the state have an unquestioned right to violate an individual's privacy—indeed, the most intimate spheres of that privacy—in the name of protecting their own agents in whatever operation they launch.

If such violations were an anomaly or part of some rogue operation, then perhaps we might view them as gross miscalculations and aberrant indiscretions of the past. But this is hardly the case, particularly when it comes to the underlying assumptions that facilitate not only the broad expansion of surveillance into the personal lives of everyday citizens but also the invasion of the privacy of individuals as part of operations in which they are not even the specific targets. The United States is by no means the sole culprit here. In the United Kingdom, for example, eight women took legal action in October 2011 against the Metropolitan Police and the Association of Chief Police Officers (ACPO) "for the harm caused by undercover officers deceiving them into long term intimate relationships."[71] In her book *Discipline and Desire*, Elise Morrison rightly notes that "within the neoliberal policing of public space, the politics of gender under surveillance has tended to receive scant attention," and perhaps one important way to counter that lack of attention is to give the case of these eight women the wide hearing that it deserves, for their specific plight has much to tell us about the politics of gender under surveillance more generally.[72]

Writing for the *Guardian*, Rob Evans and Paul Lewis reported that "most of the claimants had long-term and serious relationships with police spies, one lasting nearly six years."[73] Their case has received widespread attention in the British media in part through the efforts of the support group Police Spies Out of Lives, a nonprofit public relations organization that provides legal and financial assistance to the women who have brought their joint case against the Metropolitan Police and the ACPO. As the support group explains: the case "concerns eight women who were deceived into long-term intimate relationships by five officers who had infiltrated social and environmental justice campaigns," and "the women [specifically] assert that the actions of the Metropolitan police officers breached their human rights, subjecting them to inhumane and degrading treatment, and disrespecting their private and family life and their right to form relationships without unjustified interference by the state."[74] Since the United Kingdom does not have the established legal tradition of basic rights to privacy that U.S. courts have enshrined in their interpretations of the Constitution, the women have based their case largely on the European Convention of Human Rights, and the language of their assertions is drawn almost verbatim from Articles 3 and 8 of the Convention.

At the most basic level, this case addresses the deeply damaging effects that, as modes of performance, the deceptive practices of undercover surveillance can have, particularly with regard to their disruption of what traditionally has been viewed as the protected spaces of intimacy. As Soma and Rynerson have noted, those spaces have drawn increasing interest among legal theorists who argue that they are at the core of what "properly defines" privacy itself, and since those spaces are of fundamental significance in "developing . . . interpersonal relationships with others,"[75] their protection is, according to this line of thought, a key social and political value. In this regard, the case that the eight women have brought against the Metropolitan police touches upon the pivotal mechanisms of working democracies and of human rights more generally. But so too does this emphasis on the right to develop personal relationships "without unjustified interference by the state," particularly interference that is the product of deception, touch upon issues of performance. Indeed, it echoes the concerns articulated by Goffman a half century ago in his discussion of misrepresentation and deception in *The Presentation of Self in Everyday Life*. There Goffman considers the unsettling consequences when an "imposter's performance approximates to the real thing." The more closely it does, Goffman argues, "the more intensely we may be threatened," and, like the British women who brought the case against the Metropolitan Police, the more legitimately we may ask whether the imposter was "authorized" or "had the right to play the part he [or she] did." For Goffman, these are both legal and ethical questions. For the real threat of deception, he argues, is the profound damage it may do to our subsequent ability to extend the kind of trust in the sincerity of others' actions that is so essential to basic human relationships that Goffman calls it "sacred."[76] The threat here is of a damage that exiles the modes of performance that Goffman likens to the "real and sometimes not well rehearsed" and leaves only the "make-believe" in its wake.[77]

The magnitude of that threat was on full display recently in hearings before the Home Affairs Committee, where the women were asked to give testimony as part of a larger inquiry that was spurred on by the media attention drawn to their legal case. In that testimony one of the women reminded the committee members that "this is not about just a lying boyfriend or a boyfriend who has cheated on you. It is not even about a boyfriend who is having another relationship with somebody else. It is about a fictional character who was created by the state and funded by taxpayers' money. . . . Again, this wasn't somebody who was just good at telling lies and covering his back; he had professional training allowing him to perpetrate the deceit."[78] The content of this testimony, which alludes to something akin to state-sponsored actor's training, is compelling enough that it is easy to overlook what was perhaps the most significant aspect of the testimony as a whole: a subtle retreat into fabricated identities. The women who testified did so using pseudonyms. In a gesture reminiscent of the Dadaist Emmy Hennings, who forged identity papers to protect European intellectuals from nationalist entities hell-bent on war and destruction, the eight women literally assumed performance identities not only to protect themselves from further abuse but also to combat the performances that were orchestrated around them and that established credibility at the expense of the most intimate realms of their privacy. Speaking thus of the consequences of such deceptive invasions of privacy, one of the other women testified: "It turns your life up upside down . . . [, and] everything you thought you knew suddenly becomes unreal."[79] One can only assume that identity itself is upset in this destructive transformation. As one of the many answers to the question how does surveillance perform—how and whom does it direct—the answer that this woman provided, testifying as she did under the protective guise of a fabricated performance identity, points toward surveillance's most basic power to direct: its ability to push the willing and unwilling alike into performative constructs and make-believe—even if those constructs sometimes involve gestures and attempts to counteract or resist surveillance as a social and political force.

Rigidly structured around ritual practices and enshrined traditions, the testimony before the Home Affairs Committee and, more importantly, before the High Court, where the women have sought redress, offers a prime example of those performative constructs, and both the Home Affairs Committee and the High Court have had a significant mediating effect, among other things, on public opinion about the various forms of activism that are associated with the case itself. In this respect, they both have tended to reinforce the directing of public opinion that is a crucial aspect of the performative foundations of surveillance more generally and of the undercover operations that exploited the women in particular. The proceedings before the Home Affairs Committee and the High Court have affected not only the environmental activism that was the original target of the undercover operation but also the human rights activism that is central to the women's civil case against the Metropolitan Police. With regard to the former, that mediating effect was on display in the repeated attempts by MPs to get the women to directly or implicitly link environmentalism with the threat of terrorism and thereby provide some presumably justifiable context for the undercover operations that victimized them. Not only did the women adeptly resist these attempts in testimony, but they pointed out how the logic of such questions constructs a dubious aura of need or necessity around a tactic that rather than being an unavoidable strategy was clearly a matter of choice and a calculated decision to invade the intimate private relations of individuals who were not targets of any investigation.

Inasmuch as the circumstances surrounding that case—that is, the deliberate targeting and attempted discrediting of political activists—are manifestations of a struggle between activists and undercover operatives that spans almost five decades in the United Kingdom and the United States, the civil case is arguably tied to the legacy of the radical redefinition of theater that began in the 1960s, particularly the aspect of that legacy associated with important moments like the trial of the Chicago Eight. The connection here is one of a profound inverse symmetry: for example, the eight male defendants in the Chicago trial, on the one hand, and the eight female plaintiffs in the case against the Metropolitan Police, on the other. The more significant inversion, however, concerns the undercover operatives involved in the two cases. In the Chicago trial, the undercover operatives testified as key witnesses in the ultimately unsuccessful case against the eight political activists. In the case against the Metropolitan Police, however, attorneys for the state sought to keep police officers, who had already been exposed as undercover operatives, from giving public testimony. Indeed, they successfully petitioned the High Court to have the case removed from the public almost altogether. In an effort to reinstate a lost aura of secrecy, state attorneys filed a motion for the entire case to be moved from the High Court to a little-known secret tribunal called the Investigatory Powers Tribunal.

Although its origins are earlier, the Investigatory Powers Tribunal is primarily the product of the United Kingdom's Regulation of Investigatory Powers Act of 2000 (RIPA), and ostensibly it is designed to protect the integrity of clandestine operations while providing a venue through which to hear complaints and adjudicate "all proceedings brought against the security and intelligence services under the Human Rights Act"—particularly with regard to, among other things, "directed surveillance" and "intrusive surveillance."[80] In this respect, there was a legitimate argument to be made that the civil case brought by the eight women against the Metropolitan Police belonged before the Tribunal. But the call to bring the case before the Tribunal has been widely criticized as an effort not to protect national security but rather to stifle a gross injustice and to "hide away a potentially embarrassing court case."[81] As the Human Rights organization JUSTICE has noted in its general call for surveillance reform: "The Tribunal operates on essentially the same secret basis as its forerunners: there is no right to an oral hearing; no right to reasons for an adverse decision; no right to know the evidence put before the Tribunal by any other party; no right to cross-examine relevant witnesses; and no right of appeal or judicial review."[82] In effect, the move from the High Court to the Tribunal replicates in a new form the performative structure that victimized the women in the first place.

At one level, the logic of that replication was amply evident in the High Court's decision to grant the petition to have the case moved to the Investigatory Powers Tribunal. In the judgment that Justice Tugendhat wrote granting the petition, he not only followed the precedent of criminalizing political dissent by comparing the actions taken against the eight women to cases where undercover officers "infiltrate conspiracies to murder, rob or commit terrorist offences," but in a rather incredulous moment, he also justified his decision by citing "James Bond" as a "famous fictional example of a member of the intelligence services" who uses "relationships with women to obtain information, or access to persons or property."[83] Of course, James Bond also lives in a fantasy world where women have names like "Honeychile Rider," "Kissy Suzuki," and "Pussy Galore"—a world so sexist that even in the early 1970s the critic Mordecai Richler observed that Ian Fleming's "Bond novels are not so much sexy as they are boyishly smutty."[84] References to biased works of fiction like James Bond novels hardly offer reassurances about the kinds of deliberations that go on behind the closed doors of the Investigatory Powers Tribunal. But even these references miss the larger issue.

That issue has to do with the extent to which shifting from the

High Court to the Investigatory Tribunal replicated the paradigm not of an individual operative but of a state organization secretly assuming a deliberative role in the personal life of a law-abiding citizen. Nowhere was this replication more directly addressed than in the testimony that one of the women, who appeared under the pseudonym "Lisa," gave before the Home Affairs Committee. There she explained:

> It is not having found out that your partner was lying about who they are; it is finding out that your most personal relationship was being controlled by the state without your knowledge. There are a group of people whose names I will never know, who I will never meet, who had control over what time we spent together, who ultimately decided when my relationship was going to finish, who would have made the decisions about whether or not Alison [one of the other women] could have children in her personal relationship. All of these kinds of decisions were being made behind the scenes by a team of people who had intimate knowledge of myself and my life, and I had no idea of their existence.[85]

If anything, the move to the Investigatory Powers Tribunal increased the circle of those who have intimate knowledge of and are making fundamental decisions about—indeed, of those who are directing—"Lisa's" private affairs without her having any say in the matter, even though she was not the target of the investigation. The only thing that has changed is that "Lisa" now knows that she doesn't know who is directing her affairs—like a theatrical production team—"behind the scenes," and this brings us to two of the most crucial aspects of the civil case that the women have brought against the Metropolitan Police.

First, there is actual duration of the surveillance operations. They cover a period that extends well over a decade. Central to the entire case brought by the eight women is an undercover police operative by the name of Mark Kennedy who, within the larger span of the surveillance operations that invaded the lives of these women, was part of a seven-year operation against environmental activists, an operation that like the one targeting the Chicago Eight in the late sixties ultimately led to no successful convictions. During that period, Kennedy was known as Mark "Flash" Stone, and for six of the years before his cover was blown, he maintained a romantic relationship with the woman who in testimony before the Home Affairs Committee appeared under the alias "Lisa." Second, when "Lisa" gave testimony regarding a "team" who was directing her intimate, private affairs "behind the scenes," her comments came as part of a response to a query made by a member of the Home Affairs Committee who asked about Kennedy's work with the Association of Chief Police Officers (ACPO), which from 2006 to 2010 had control of operations like the one involving Kennedy and which, as the committee member was quick to point out, is a private company. More than anything else, the depth of Kennedy's involvement in "Lisa's" private life—six years out of a seven-year operation that ultimately led to no successful convictions—underscores the extent to which his operation was more about directing than about detection and prosecution of crime. But given his affiliation with the ACPO, so too was this directing about more than state invasion of an individual's privacy. It was about the state hiring private companies to invade "Lisa's" privacy, and about the state giving private companies leave to direct her private affairs. Inasmuch as this directing involved a team "behind the scenes," questions like those coming from the Home Affairs Committee member about the ACPO indicate that "Lisa" was not the only one who was uncertain about the team's identity. If the Home Affairs Committee itself was somewhat uncertain about that identity, this was because the team directing "behind the scenes" proved to be a moving target, blurring the boundaries separating operations conducted by the state and those conducted by private enterprise.

From Gray Intelligence to Gray Surveillance: The Case of Mark Kennedy

In the midst of that kind of blurring, the significant differences between the United States and the United Kingdom on issues of privacy and surveillance begin to disappear. Soma and Rynerson have noted, for example, that "in the United States, privacy is most often equated with liberty from an intrusive government" and that "traditionally, Americans have been, if not necessarily trusting of the business sector, then at least not highly skeptical of it; instead reserving their deepest distrust for the institutions of government, and particularly the Federal government."[86] With regard to matters of surveillance, this lack of skepticism is deeply puzzling since it so readily plays into an established tradition of farming intelligence tasks out to private companies in ways that ultimately circumvent democratic oversight and that twist the priorities of national security and law enforcement into profit-driven business models. In this respect, the United States and the United Kingdom have followed similar neoliberal trajectories, and the ambiguous status of agents like Mark Kennedy is a very subtle reminder that the use of undercover operatives has a long history in something akin to what Eveline Lubbers, echoing the Dutch scholar A. B. Hoogenboom, calls "gray intelligence."

For Hoogenboom and Lubbers "gray" denotes an operative realm devoid of clarity, where black-and-white distinctions between the interests of the state and of private enterprise have faded into a common indistinguishable blur. According to Lubbers, gray intelligence begins with the state privileging corporate priorities in its use of intelligence services, the kind of privileging, for example, that was ideologically central to COINTELPRO under Hoover's guidance. Such priorities, Lubbers argues, gradually allow for a legal privatization of intelligence services "with a wide variety of private entrepreneurs"[87] who, commissioned to do the government's intelligence work, shape it increasingly after their own image, after their own ideological assumptions, and, above all, after business models that will sustain their own industry and profits.

Whereas Lubbers sees this process as one dealing primarily with the collection of intelligence, I would suggest that by slightly modifying her basic concept of gray intelligence to "gray surveillance," we move toward a more accurate depiction of how the legacy of COINTELPRO's undercover operatives and their unacknowledged role in the 1960s' radical redefinition of theater continues to shape surveillance today. What actually endures from that legacy—despite all of the profound advancements in the technologies of surveillance since the 1960s—are surveillance's performative foundations. While the notion of gray surveillance highlights the increasingly dominant role that private business and corporate interests now play in conducting surveillance as a marketable commodity, this role has further grounded surveillance in models of performance and calculated strategies of directing. At the most basic level, gray surveillance suggests not so much a gathering of intelligence as a blurring of state and private interests in the use of intelligence resources to disrupt, misdirect, and contain activists or any other group or organization running counter to corporate priorities. It also suggests a gradual privatization of such resources: a legal outsourcing of the ability to direct—on both large and small scales—into private corporate hands.

If gray surveillance means the ability to repress, manipulate, and control in the ostensible name of the state, so too does it mean the ability to do so in the practical pursuit and development of a sustainable business model. More than anything else, it is this pursuit of a sustainable business model that in the late twentieth and early twenty-first centuries has cultivated and strengthened surveillance as a mode of directing. In many respects, the question "How does surveillance perform?" is the quintessential question central to any business model, and the answer has everything to do with calculated acts of directing. In business terms, this question of performance is as much about how well the models of surveillance supply clients with a desirable commodity as it is about how much surveillance as a commodity generates continued demand for the services that the surveillance industry provides. Indeed, if there is motivation to instigate and direct rather than to investigate and detect—a desire to neutralize, disrupt, and vilify ideological competition and political opposition rather than to identify and prosecute crime—then it is to be found in the hands of private enterprise, which has as much interest in neutralizing anything that looks like viable anticorporate activism as it does in convincing government entities of the urgent need to continue subscribing to the packaged vision of society that private surveillance companies market.

In the post-9/11 era, there are an alarming number of examples of what I am calling "gray surveillance." They are a product of the grotesque expansion of what scholars like Kristie Ball and Laureen Snider have described as "the surveillance industrial complex."[88] But Ball and Snider are by no means alone in this description. Long before Edward Snowden became a household name, for example, Tim Shorrock had written eloquently on "the intelligence industrial complex" and on the profound influence that the company Booz Allen Hamilton (which was Snowden's employer) has gained in government affairs during the post-9/11 era.[89] Reminiscent as such phrases are of Eisenhower's famous admonishment about the "military industrial complex" in the last speech of his presidency,[90] allusions to "the surveillance industrial complex" and "the intelligence industrial complex" offer a trenchant characterization of the dangerous expansiveness of private surveillance industries—a characterization that implicitly equates the consequences of the privatization of surveillance technologies and techniques with the threat to democracy that Eisenhower warned would be the result of private corporations having an increasing stake in decisions about military readiness and defense.

With regard to surveillance, those dangers come big and small, and though "gray surveillance" aptly characterizes the operations of large corporate entities like Booz Allen Hamilton, so too does it characterize undercover operations like the one that involved Mark Kennedy. At one level, the case of Mark Kennedy is significant because it unfolded outside of the long shadow cast by the events of September 11, 2001, in the United States and the events of July 7, 2005, in the United Kingdom, and in this respect it offers a counter to the notion that the blurring of state and corporate interests into gray surveillance resulted from the need to respond quickly to the pressing threats of global terrorism. Although Mark Kennedy's seven-year stint as an undercover operative in the British and European environmental activist movements[91] began around 2002, it had little or nothing to do with the events of September 11. Indeed, when his cover was blown in late 2010, his exposure created a political scandal in the United Kingdom that ultimately led to the outing not only of eleven other undercover police officers but also of a four-decade-old program in the United Kingdom that was comparable to COINTELPRO in the United States. That program goes all the back to the antiwar demonstrations in the 1960s. It was called the Special Demonstration Squad, and, ironically, it was given the acronym SDS. But this was no British variant of the Students for a Democratic Society. The SDS was formed by Conrad Hepworth Dixon in September 1968 after the Grosvenor Square demonstration in March of that year, and was specifically tasked with infiltrating "left-wing political groups."[92] As Rob Evans and Paul Lewis have noted, however, what began as an undercover policing of antiwar activists in the late 1960s "became an enduring and permanent programme of state espionage, directed at political activists."[93]

As the case of Mark Kennedy illustrates, by the early 2000s that program had taken on an increasingly global scale. For Kennedy was active not only in the United Kingdom, but also on the European continent as well. Indeed, during the seven-year operation built around his undercover identity—an operation that ultimately cost British tax payers over £1.75 million[94] and that led to no criminal convictions—Kennedy traveled to some twenty-two different countries, "joining protests from Iceland to Poland."[95] What is telling about the global quality of Kennedy's work is not the suggestion that he embodied the questionable global concerns of the British government but rather that from the very beginning, his work exemplified an evolution into "the gray" of the kind of undercover operations that began with COINTELPRO and the United Kingdom's SDS.

In 2010, not only was Kennedy exposed as an undercover operative and as an agent provocateur who facilitated the arrest of some twenty-six activists on conspiracy charges after helping them devise a plan "to break into Ratcliffe-on-Sour power station" in 2009—an exposure that ultimately led to a collapse of the case against them[96]—but at the time of his exposure he had already resigned from the Metropolitan Police, had set up a private consulting firm called Troka Limited, and was working as a consultant for a lucrative private surveillance company called Global Open. Global Open was set up by a former undercover operative by the name of Rod Leeming, and it is, as Rob Evans and Paul Lewis have noted, merely one of a number of private detective agencies like the Inkerman Group, Vericola and, C2i International that make money by spying for large corporations and that move into what we might call the "grayest of the gray" since they are "completely uncontrolled and unrestrained" in their activities[97] by democratic oversight. While they take government contracts, they are governed by market revenue and serve corporate interests.

Yet even before Kennedy resigned from the Metropolitan Police and became a consultant, his work was shaded in gray. Eveline Lubbers has noted that it has never been entirely clear for whom Kennedy was working in the first place. During his seven-year assignment, Kennedy was tasked in part under the auspices of the Association of Chief Police Officers, which as members of the Home Affairs Committee have pointed out is not under government control. Lubbers has noted that the ACPO "is not a public body of any sort" but rather "a limited company" not only "sharing its data with clients such as energy companies running power plants and airline companies involved in the expansion of airports" but also "selling information from the Police national computer for up to £70—even though it pays just 60p to access those details."[98] Kennedy's involvement with ACPO laid the foundation for his later work with Global Open, and ultimately for the employment he found when after his exposure he fled to the United States. Despite the numerous public statements that he has made since his exposure about the enduring sense of allegiance he feels for the activists he deceived and betrayed, Kennedy almost immediately found gainful employment in the United States in the private sector, drawing upon his vast experience as an undercover operative in activist circles. As of the summer of 2012, Kennedy was working for the private company Densus Group, "advising corporations on how to deal with the 'threat' of political activists."[99]

In their book *Undercover*, Rob Evans and Paul Lewis suggest that this kind of blurring between public and private interests is typical, and they cite senior police officers who claim "that 'without question' there are more corporate spies posing as protesters" than there are "those run by police." Those corporate spies are free from the legal constraints, protocols, and codes of conduct that govern police activity.[100] Corporate undercover operatives are, in short, freer to improvise and to take liberties with those with whom they have contact. In many respects, this is the unspoken objective of gray surveillance. Outsourcing alleviates the burdensome task of oversight, and it changes the fundamental dynamics of undercover surveillance as a theater of operations. It changes how surveillance performs. But the issue here, as has been the issue throughout this chapter, is not merely the privatization of undercover surveillance in corporate hands but what undercover operations as modes of surveillance tell us about how surveillance performs more generally. Inasmuch as the privatization of undercover surveillance is merely one example of a more general privatization of surveillance techniques and technologies, that telling bears but a faint echo of the radical redefinition of theater that took place on the streets of the 1960s. It is a telling of the largely uncontrolled and unrestrained expansion of surveillance beyond the streets and into all aspects of society. It is a telling of an expanding capability of surveillance's ability to perform and, above all, of its ability to direct.

This Private Island: The Modern Ban-opticon from Camp X-Ray to Private Prisons

> Where there is no publicity there is no justice.
> —Jeremy Bentham

Turning a Blind Eye on Camp X-Ray: Preliminary Reflections on Surveillance and a Postpanoptic Paradigm

Over a decade has passed since Karen Greenberg and Joshua Dratel edited the thirteen-hundred-page collection of memos and reports by U.S. officials that was then published under the title *The Torture Papers: The Road to Abu Ghraib*. The memos and reports published in *The Torture Papers* provided what the *New York Times* called "a damning paper trail" of calculated circumventions of U.S. judicial practices and of what subsequent legal historians have characterized as perhaps the single biggest "rerouting" of the fundamental principles of due process to be found in the history of modern Western democracies. Indeed, keeping with this notion of roads, trails, and reroutings, Joshua Dratel introduced the legal documents in *The Torture Papers* by writing that "while the proverbial road to hell is paved with good intentions, the internal government memos collected in this publication demonstrate that the path to the purgatory that is Guantanamo Bay, or Abu Ghraib, has been paved with decidedly bad intentions."[1] As a point of departure for this chapter, I would like to place the legal "paper trail" of *The Torture Papers* within the context of a larger discussion of surveillance and performance that focuses on three primary areas of consideration.

First of all, I want to explore how the political and juridical backdrop to the detention facilities at Guantánamo Bay, Cuba, might offer valuable lessons about surveillance and surveillance technologies more generally since the prison there is in a very literal sense a constructed maximum surveillance environment, one that was consciously designed to elude legal oversight. Second, I want to examine the ways in which such environments restructure, indeed "reroute," the power dynamics of performance itself. To that end, I want to juxtapose my introductory comments on the legal obfuscations documented in *The Torture Papers* with a brief analysis of Victoria Brittain and Gillian Slovo's *Guantanamo: Honor Bound to Defend Freedom*. My interest here is not so much in this work as a post-9/11 drama of the so-called war on terror but rather in how this piece, relying as it does on the conventional divide between spectators and performers, implicitly reminds us that Guantánamo's "Camp X-Ray"[2]—both in terms of its physical location and in terms of its conceptual structures—is a site of complex theatricality that can tell us much about the complex theatrical underpinnings of surveillance itself. Third, I want to bridge that notion of theatricality with the very literal sense of theatricality that is written into Athol Fugard, John Kani, and Winston Ntshona's antiapartheid drama *The Island* and its play within a play. I argue *The Island* not only presents us with an important compliment to Slovo and Brittain's drama but does so in such a way as to highlight the underlying racial politics that give structure and form to the modern surveillance society and that reach all the way from the coast of the African continent and apartheid to the U.S. mainland, to its prison industrial complex and to the American surveillant assemblages that Michelle Alexander has called "the new Jim Crow." All of this is geared toward gaining a critical understanding of the complex mechanisms of the surveillance society that are evident in the theatricality of maximum surveillance environments.

At the most basic level, the case for that theatricality is not difficult to make. It is grounded in what—with the possible exception of George Orwell's image of Big Brother—is arguably the single most dominant metaphor of surveillance among cultural critics: the theoretical model of the panopticon that in the latter half of the twentieth century Michel Foucault popularized in his reading of the utilitarian philosopher Jeremy Bentham. Bentham is a particularly appropriate point of reference here because, conceptually at least, Guantánamo's "Camp X-Ray" owes a profound debt to the pervasive sense of visibility that Bentham offered as a mechanism of control in modern prisons. As I noted earlier in chapter 1, which focused on the theological metaphors that dominate the discourse of surveillance, prisoners in Bentham's proposed model would be housed in illuminated cells circling a central tower in which a single guard, whom the prisoners could not see, would be able to monitor each individual prisoner's actions. Not knowing when exactly the guard was monitoring them or whether intervention and subsequent punishment might occur, the prisoners—so the theory goes—would modify their behavior and conform to prison rules.

Foucault famously saw in Bentham's proposal a model of surveillance in society more generally—as did his contemporary, the psychoanalyst Jacques-Alain Miller, who is perhaps best known as the close associate of Jacques Lacan. In fact, it is difficult to determine who discovered Bentham first. In the same year that Foucault published *Discipline and Punish* (1975), Miller published his own discussion of Bentham entitled "Jeremy Bentham's Panoptic Device." There, Miller argues that "the Panopticon is not a prison. It is a general principle of construction, the polyvalent apparatus of surveillance." Implicitly recognizing that Bentham had modeled his idea of a circular prison after the plans that his brother Samuel Bentham had developed for a Russian textile mill, Miller notes that "the Panoptic configuration could be used for prisons as well as schools, for factories and asylums, for hospitals and workhouses. It has no unique application," he argues, and it is "designed to house involuntary, unwilling, or constrained inhabitants."[3] I want to return to Miller's essay momentarily since his reading of Bentham supplements Foucault's in a way that underscores what is arguably Bentham's most significant relevance for surveillance studies today. But first, it is worth noting that both Miller and Foucault highlighted the inherently theatrical aspect of the panopticon and thus, by extension, of surveillance itself. "By the effect of backlighting," Foucault notes, "one can observe from the tower, standing out precisely against the light, the small captive shadows in the cells of the periphery. They are like so many cages, *so many small theatres*, in which each actor is alone, perfectly individualized and constantly visible."[4] Miller, taking a more comprehensive view of the panopticon, called it "a veritable theatre of punishment," and, quoting Bentham himself, Miller added that "it offers the spectator" what Bentham called "a perpetual and perpetually interesting drama, in which the obnoxious characters shall, *in specie*, at any rate, be exposed to instructive ignominy."[5]

This acknowledgment of theatricality is richly provocative, but subsequent scholars in the field of surveillance studies have tended to be quite critical of Foucault's sweeping extension of a late eighteenth-century conceptual model of surveillance into twentieth- and twenty-first-century technological realities. In their influential essay "The Surveillant Assemblage," for example, Kevin Haggerty and Richard Ericson express dismay that "Foucault fails to directly engage contemporary developments in surveillance technology, focusing instead on transformations to eighteenth and nineteenth century total institutions."[6] More recently, Zygmunt Bauman and David Lyon have advocated a postpanoptic perspective in their book *Liquid Surveillance*. Drawing upon the work of Didier Bigo, they embrace the notion of what Bigo calls the "ban-opticon," which, rather than having a central all-seeing authority (as symbolized in Bentham's guard tower), operates through diffuse "state and corporate entities," sorting individuals into various social categories. Instead of producing models of quarantine, detention, or confinement, such sorting, they argue, determines whether one "is welcome or not," and whether one is granted access to or is banned from various locations and activities, or from various rights and privileges. Ultimately, the ban-opticon determines, according to Bauman and Lyon, whether one is to be "excluded not just from a given nation-state"—as in being denied passage to cross a national border—but also whether one is to be excluded "from a rather amorphous and not unified cluster of global powers."[7] Whereas the panopticon supposedly cultivates a mentality of subservience and submission because inmates, and ultimately citizens as well, are, to follow Foucault, "caught up in a power situation of which they are themselves the bearers,"[8] the ban-opticon renders the cultivation of such mentalities largely obsolete, functioning instead—through the social sorting that it performs—as a gatekeeper not to keep prisoners confined but to keep the disenfranchised and unauthorized out of privileged spheres where wealth, power, and authority are consolidated and reconfigured into radically new and "postdemocratic" political orders.

"Amorphous and not unified" though those new political orders may be, they bear a remarkable resemblance to a technologically supercharged version of what roughly a decade ago the sociologist Colin Crouch described as "postdemocracy"—that historical moment "when powerful minority interests have become far more active than the mass of ordinary people in making the political system work for them"[9] and when "the rise of global capitalism . . . [produces] a self-referential class more concerned with forging links with wealthy business interests than with pursuing political programs which meet the concerns of ordinary people."[10] Crouch may very well be correct in claiming that among the masses a sense of "boredom, frustration and disillusion"[11] with the political system contributed to the rise of postdemocracy, but whatever the causes of this political shift, I want to connect some important historical dots and suggest that the post-panoptic regimes of surveillance arguably sustain that postdemocratic order. While I will only touch upon Crouch's argument here, I will return to it at great length in the final chapter of this book, where I explore the future of theater and performance in an era of postdemocracy and of the rise of the surveillance society.

While the social sorting and hierarchies of exclusion that the banopticon produces are supported by profound advances in surveillance technologies like biometrics, dataveillance, digital processing, and sophisticated global tracking software, it is the conceptual shift from the panoptic to the ban-optic that marks the fundamental difference between the eighteenth-century surveillance complex conceptualized in Bentham's panopticon and the twenty-first-century surveillance complex exemplified in Guantánamo's Camp X-Ray. One can and must address the horrible acts of injustice that, in the ostensible name of democracy, the United States continues to perpetrate at Guantánamo.[12] Indeed, beyond the plight of individual prisoners like Jamal Al-Harith (released 2004), Bisher Al-Rawi (released 2007), Moazzam Begg (released 2005), and Ruhal Ahmed (released 2004), the purpose of Brittain and Slovo's documentary drama *Guantanamo: Honor Bound to Defend Freedom* is largely that of issuing a call for justice for all of the prisoners held there (of the 779 prisoners brought to Guantánamo 41 remain as of February 2017).[13] But those acts of injustice, morally outrageous though they may be, are merely symptoms of the much deeper political malaise that Guantánamo itself exemplifies as one particularly revealing instance of a ban-optic site of discipline and surveillance. While 9/11 may have exacerbated that malaise, it did not cause it, and the significance of Guantánamo—even more a decade after its founding—is that it pulls back the curtain on the malaise, ironically exposing the U.S. government's efforts to remove from scrutiny its own crass rejection of democratic oversight and its defiance of international laws and conventions.

So profound are the shifts in political power exemplified at Guantánamo that it is far from unrealistic to speculate that future critics and historians are likely to theorize Guantánamo's Camp X-Ray much as Foucault and Miller did the panopticon. Even now we might constructively ask what that theory might entail. In answering that question, one could easily begin with a simple point of contrast between the legal obfuscations that gave rise to Guantánamo and the basic principles of transparency that critics and historians have tended to overlook in Foucault's and Miller's readings of Bentham. With regard to Guantánamo, some basic sense of the legal obfuscations can be garnered from the corpus of Karen Greenberg's work. Four years after publishing *The Torture Papers*, she reported in 2009 that from Camp X-Ray's beginnings in January 2002 the original commanding officer, Marine Brig. Gen. Michael Lehnert, who had been assigned the task of establishing the detention camp at Guantánamo, worried about "the policy vacuum into which he and his troops had been thrown." Although "he had been told by the Joint Chiefs of Staff that the Geneva Conventions would not technically apply to his mission . . . [and that] he was to act in a manner 'consistent with' the conventions (as the mantra went) but not to feel bound by them,"[14] there was disturbing lack of clarity about how to handle the prisoners. What Lehnert was to learn after essentially losing his command a month later to Maj. Gen. Michael Dunlavey, an Army Reserve officer and interrogation specialist who was specifically sent to Guantánamo to do Donald Rumsfeld's bidding, was that the policy vacuum was not the problem. It was in fact the point of Guantánamo. Over a decade later, there is little debate today regarding the widespread understanding that Guantánamo was established to keep detainees in a perpetual state of legal limbo and to keep them out of reach of the U.S. judicial system. But what, in this regard, Guantánamo tells us about surveillance has less to do with the relocation of prisoners to a site where, for all intents and purposes, they have been banished beyond the reach of normal legal precepts than it does with placing them at a site from which the public itself has been banned—a site where, as a consequence, prison officials can make decisions and act without the hindrance of meaningful democratic oversight. If theater is a place where people go to see and be seen, the novel theatrical construct of Guantánamo was thus to create a site where the spectator's gaze remains unchecked. It created a site, in other words, where no one watches the watchers. Therein lies the surprising point of contrast.

If Guantánamo was consciously designed to elude basic structures of accountability, nothing of the sort is true of the panopticon. The guard in Bentham's tower, while not visible to the prisoners, is not exempt from oversight. This may come as a bit of a revelation to many since in popular discussion, Foucault's theoretical concerns with the mechanisms of authority and power generated by the panopticon are compelling enough that they often eclipse what he and Miller both identify as Bentham's crucial safeguard against the panopticon degenerating into a mechanism of tyranny: a visibility that moves in all directions. That visibility is a key aspect of Bentham's social political philosophy. In "Constitutional Code," for example, Bentham famously wrote: "In the darkness of secrecy, sinister interest and evil in every shape, have full swing. Only in proportion as publicity has place can any of the checks, applicable to judicial injustice, operate. Where there is no publicity there is no justice."[15] It should come as no surprise then to find a similar mechanism of public oversight structured into Bentham's conception of the panopticon. As Miller notes: "In the Benthamic system it is crucial to know who watches the guards. Here is the answer: the eye of the public will watch over the inner eye." For the central tower in the panopticon design is open to the public, and in Bentham's conception, the public could visit the panopticon at will. The understanding was not only that members of the public would gain valuable instruction from observing the prisoners paying the consequence of their crimes, but that the presence of the public would have the added benefit of hindering abuse in the prison itself. Miller thus continues: "The visitor, while learning from the spectacle, will also inspect, control, and organize. In this way Benthamic space becomes totally 'panoptic': in turn, the invisible surveillance reintegrates visibility; the watcher becomes the subject of surveillance."[16] Miller is not alone in this reading of Bentham. Referring to public access to the panopticon,

Foucault himself argues in a seldom cited passage that "there is no risk, therefore, that the increase of power created by the panoptic machine may degenerate into tyranny; the disciplinary mechanism will be democratically controlled, since it will be constantly accessible 'to the great tribunal committee of the world.'"[17]

Despite its popularized image, Bentham's panopticon, as a model of surveillance, was thus grounded in a mechanism of oversight, checks, and balance. In short, the panopticon required transparency. This is not to suggest that such mechanisms are exempt from manipulation, particularly in an increasingly mediatized society. David Lyon has cautioned, for example, that "in contexts where opportunities for media manipulation are rife, there may be an appearance of openness, transparency and communication, but no accountability worth the name,"[18] and perhaps this is one way in which we have entered into a postpanopticon society. But there are more profound ways as well. In Bentham's conception, accountability walks hand in hand with publicity, and there is but a small distinction between his sense that "where there is no publicity there is no justice" and his sense that public access to the panopticon would make its administrators accountable before "the great tribunal committee of the world."

Gitmo' Sight Unseen: Imagining What Surveillance Deems Not to Share

Keeping this notion of "the great tribunal committee of the world" in mind, one begins to understand what, seen through the lens of competing theories of surveillance, Brittain and Slovo's work *Guantanamo: Honor Bound to Defend Freedom* attempts and what it ultimately reveals. As Ryan Claycomb noted in his review of the 2005 Studio Theatre production of the play in Washington, DC, Brittain and Slovo's piece "revolves around the case of four British Muslims, all detained at the US naval base at Guantanamo Bay," and "using testimony from family members and attorneys, public statements from officials, and letters and testimony from the detainees, the play deploys the story of these men to interrogate US detention practices in Cuba."[19] Against the backdrop of this theatrical interrogation, Brittain and Slovo's work points first and foremost toward a notion of the ban-opticon that is far more insidious than Lyon or Bauman suggest in their echoes of Didier's theories.

It points toward a notion of the ban-opticon where it is not merely individuals who may face bans but also the public at large. In this regard, Brittain and Slovo, metaphorically at least, are knocking at a tower door that has been locked in history. That is to say, through the medium of theater, their work would symbolically reinstate a panoptic model of surveillance upon a postpanoptic—indeed, what arguably is a broad ban-optic—historical event. It casts the public—that "great tribunal committee of the world"—in the form of a theater audience and places them precisely at the site where they have been banned: inside Guantánamo, where they might provide what Bentham called "the checks, applicable to judicial injustice." As politically laudable as I find this gesture, at one level it is immersed in a moment of profound nostalgia—one that expresses a longing for a moment in the evolution of surveillance and democracy that it cannot actually recover. In short, *Guantanamo: Honor Bound to Defend Freedom* (re)stages a historical moment prior to the advent of the postpanoptic and of the postdemocratic, and it imposes it upon a site that is the very quintessence of a ban-optic surveillance apparatus.

But if a work like *Guantanamo: Honor Bound to Defend Freedom* succumbs to a moment of nostalgia by symbolically imposing a panoptic model of surveillance over a ban-optic historical moment, how then is the actual detention camp at Guantánamo Bay to be understood as a postpanoptic site of surveillance? What theoretical model does it resemble as a powerful example of the structures of surveillance in a ban-optic society? Before moving into a much broader discussion that reaches well beyond Guantánamo and the war on terrorism and that begins to explore the rich conceptual significance of island prisons as a metaphor for contemporary structures of surveillance, I would like to suggest that some conceptual structures of that theoretical model were already beginning to take shape even as Brittain and Slovo's play premiered at London's Tricycle Theatre in 2004. Ironically enough, those structures were held together by a common foundation in an infamous study conducted in 1971 in the basement of the Stanford University Psychology Department by the American psychologist Philip Zimbardo, a study known as the Stanford Prison Experiment.

In fact, at roughly the same time that Brittain and Slovo's play premiered, a rather astonishing renaissance of interest in Zimbardo's study was occurring in the arts, in the social sciences, and in mainstream politics. It is worth considering at some length the insights and, ultimately, the limitations that Zimbardo's study thus has for our understanding of the logic of surveillance more generally. In the final sections of this chapter, I will explore the blind spots of Zimbardo's study in some detail since they are of major significance to our understanding of the role that politics and race play in the differing structural forms that surveillance takes in society. But before pushing into those blind spots, I want to examine carefully the insights that Zimbardo's study offers because they too are significant, particularly with regard to sites like Guantánamo. Amid growing concerns about abuse of prisoners at Guantánamo, Abu Ghraib, and numerous "black sites" by the U.S. military, the CIA, and private security personnel contracted by the U.S. government, the broad spectrum of renewed interest in Zimbardo's study would be hard to dismiss as a matter of coincidence. While each of its currents—in the arts, the social sciences, and politics—carried a specific nuance that has particular relevance for Guantánamo as a site of detention, taken together they form the structures for what provides a rough understanding of Guantánamo as a ban-optic site of surveillance.

The Stanford Prison Experiment: Surveillance Adventures in Art, Media, and the Social Sciences

Although the political interest in Zimbardo's study may provide the decisive aspect of that understanding, it is worth pausing momentarily to see how the interest in Zimbardo's study in the arts and social sciences provide a critical frame for the discussion of Stanford Prison Experiment among mainstream politicians. Consider, for example, that in the year following the premier of *Guantanamo: Honor Bound to Defend Freedom*, the Polish artist Artur Zmijewski released his film *Repetition*—a film that bore remarkable similarity to a 2002 BBC series produced by Steve Reicher and Alex Haslam at the University of Exeter. That series was officially titled *The Experiment*, but it has subsequently come to be known as the BBC Prison Study since it was broadcast on BBC television. In many respects, the rough theoretical outlines for conceptualizing Guantánamo as postpanoptic model can be found in way that these two pieces speak to one another not only in terms of their shared historical debts and similarities but also in terms of their profound differences. Although Zmijewski conceptualized his piece as a work of art and Reicher and Haslam conceptualized theirs as a type of social scientific research suitable for the BBC's broad public audience, both pieces took a marked departure from the conventional structures of drama and theater used in the Tricycle Theatre production of *Guantanamo*. Unlike Brittain and Slovo's documentary drama, which relied on scripted performances of professional actors, Zmijewski's *Repetition* and Reicher and Haslam's *The Experiment* focused on the improvisational behavior of "volunteers" who participated in constructed environments where for approximately one week they assumed the loosely defined roles either of prisoners or of prison guards. Although technically speaking *The Experiment*—at least according to Reicher and Haslam—was not intended "to simulate a prison (which . . . would have been impossible on ethical and practical grounds) but rather to create an institution that in many ways resembled a prison,"[20] both *Repetition* and *The Experiment* purported to explore how average individuals negotiate their way through constructed authoritarian environments: examining, for example, how far individuals who have been granted substantial authority over others will push that authority to its extremes and become abusive, or conversely how far individuals will ultimately submit to another person's authority even if submission entails personal humiliation, psychological harassment, or physical abuse. In this respect, both films were centered on conditions that either consciously mirrored or were directly inspired by the Stanford Prison Experiment conducted by Zimbardo back in 1971.[21]

In that original experiment, Zimbardo took twenty-four student volunteers, constructed a makeshift prison in the basement of the Stanford Psychology Department, randomly divided the students into prisoner and guards, and over the course of six days observed with amazement as those cast as prisoners became increasingly submissive and as those cast as guards became increasingly sadistic. All were deeply traumatized by the experience. The Stanford Prison Experiment gained infamous notoriety because it so quickly spun out of control, devolving into acts of such brutality that, despite a planned two-week duration for the experiment, Zimbardo had to end it after only six days. Indeed, the Stanford Prison Experiment's disastrous results were in part responsible for a fundamental ethical and legal reassessment of permissible use of human subjects in social scientific experiments. Today, as Anthony Downey has noted: "On ethical grounds, the original Stanford Prison Experiment would be not only inadmissible . . . but legally unperformable under the guidelines set out by the American Psychological Associate Ethics Code, the Canadian Code of Conduct for Research Involving Humans, and the Belmont Report."[22] Downey's reference to these guidelines provides the basis for a scathing critique of Zmijewski's *Repetition*. Looking at the consequences of Zimbardo's original experiment, Downey questions the ethical justification for Zmijewski's restaging of an experiment in the name of "art" that would otherwise be ethically and "legally unperformable" in the name of research. This is understandable. Given the instances of abuse that also surface in Zmijewski's film, one would be hard pressed not to agree with Downey, who not only questions Zmijewski's rather manipulative recruitment of unemployed Polish young men for his project—whom he bullied into remaining with the project by refusing to pay them if they left—but also the ethical motivations behind the project as a whole.[23]

Beyond the limited question of art and ethics, however, there is the broader issue that is raised in the disturbing precedent set by Zimbardo's experiment, a precedent echoed in Zmijewski's *Repetition*. That precedent has more to do with the constructed context for the role-playing that the original experiment and its later restaging demanded of their participants than it does with the deeply disturbing zealousness with which some of the participants played those roles. Critics have questioned both the original experiment and Zmijewski's restaging of it because of the adverse effects it had across the spectrum of participants, regardless of whether the participants were cast as prisoners or prison guards. In discussions of Zimbardo's study and Zmijewski's film, the question of ethics has thus consistently centered not on the abuse of prisoners per se—ethically and morally repugnant though that abuse certainly is. It has focused instead on the creation of a context and environment where such abuse is facilitated, cultivated, and even implicitly encouraged. In the rationale that they offer for using the BBC Prison Study as an alternative to Zimbardo's original study, for example, Reicher and Haslam link this question to the issue of leadership. Not only do they question whether the abusiveness of those whom Zimbardo cast in the role of guards actually resulted "as Zimbardo and his colleagues claim" because of what he has often described as "their 'natural' acceptance of role requirements." Reicher and Haslam also suggest that a better explanation for the abusiveness was "the leadership provided by the experimenters."[24]

Perhaps the "absence" of clear leadership would be a more accurate description. In contrast to Reicher and Haslam, Downey suggests, for example, that the astonishing abuse that guards "meted out to the prisoners [in the Stanford experiment] was largely due to the lack of any clear guidelines from Zimbardo as to how detainees should be treated."[25] But in many respects, the difference between Reicher and Haslam's argument and Downey's is merely semantic. For often the lack of clear guidelines serves an identifiable purpose. If "leadership" led to the collapse of the Stanford Prison Experiment, it was the kind of leadership that draws participants in by creating a vacuum that the leaders, without ever saying so, then expect the participants to fill. Ultimately, Zimbardo concedes this very point in his 2007 book *The Lucifer Effect*. There he admits:

> The design of the Stanford Prison Experiment made it evident that initially our guards were "good apples," some of whom became soured over time by powerful situational forces. In addition, I later realized that it was I, along with my research team, who was responsible for the system that made that situation work so effectively and so destructively. We failed to provide adequate top-down constraints to prevent prisoner abuse, and we set an agenda and procedures that encouraged a process of dehumanization and deindividuation that stimulated guards to act in creatively evil ways.[26]

There is a fine line, of course, separating a failure to provide "adequate top-down constraints" and implicitly setting "an agenda and procedures" that encourage abuse. But however fine the line might be, it is one that points directly from the Stanford Prison Experiment to Guantánamo.

A Wink and a Nod, and the "Exceptionalism of Power"

The clarity of that line resides in the force of the knowing wink and the authoritative nod—in other words, in the kind of coercive political equivocation that demands results while capitalizing on the privileged kind of power capable of circumventing accountability by never needing to state openly what is clearly implied and obviously expected. This is the kind of equivocal coerciveness that Zmijewski applied with the power of the purse in *Repetition* when he drafted unemployed and economically desperate young Polish men for his "artistic" project and demanded that they "perform" or forgo payment. The fear of economic reprisal made for "enthusiastic" performers among guards and prisoners alike. But so too is it the kind of equivocation that possesses an eerie similarity to the situation at Guantánamo and Abu Ghraib where the postpanoptic and the post-democratic converged in a defining instance of the ban-opticon. And that is precisely the point. Some thirty years after the original Stanford experiment, Zimbardo and other scholars have drawn direct parallels between the experiment's lack of guidelines and its disturbing consequences, on the one hand, and "the absence of any Standard Operating Procedure (S.O.P.) at Abu Ghraib"[27] and the horrific human rights violations perpetrated there by U.S. soldiers, CIA operatives, and private contractors, on the other. There leadership was conducted by insinuation, implication, and equivocation. It was conducted in hidden shadows rather than in the light of day where it might be subject to oversight and control and where its accountability to Bentham's "great tribunal committee of the world" might have checked the downward spiral into horrible violations of human rights. This is the model of surveillance exemplified at Guantánamo.

The surveillance regimes at Guantánamo epitomize the kind of leadership that is reprehensible because it pursues an agenda that allows its instigators and sponsors to hide behind the false aura of transparency while their subordinates, whom they pressure into committing despicable acts, take the fall and are held accountable for a script they did not author, a script in which they merely played exacting but ultimately minor roles. I would hardly be the first to note the theatricality in this arrangement. Zimbardo himself has recently suggested that "many key players in the drama at Abu Ghraib—the directors, scriptwriters, and stage managers who made this tragic play possible" sought to remain "behind the dark screen of concealment" while the MPs (military police) were left to improvise and "were merely bit actors, 'seven characters[28] in search of an author,' or a director."[29] But it is not as if the authors and directors could not be found. They simply could not be held accountable because they were careful to use structures of power and command that reflect what Didier Bigo, in characterizing the key operative elements of the ban-opticon more generally, calls the "exceptionalism of power."[30]

This exceptionalism is embedded in a deceptive theatrical sleight of hand and in a moment of plausible deniability. In a very literal sense, it represents the near perfection of a political theatricality that, to echo David Lyon's comments from earlier, gives the "appearance of openness, transparency and communication, but [ultimately provides] no accountability worth the name." Bigo argues that "to speak of the exceptionalism of power" is to speak of more than the special measures and emergency acts that suspend the normal checks and balances of power in exceptional cases or circumstances. It is to speak of how such acts "symbolically" legitimize the effects of extraordinary laws and measures—effects like sanctioning torture and other human rights violations—giving them precedent and a basis to become ordinary and commonplace. Since such measures necessitate a suspension of the normal mechanisms of checks and balances, they also give precedent to the suspension of oversight, making it commonplace and acceptable as well. Powerful interests—sometimes government interests and sometimes contracted private corporate interests—are thus in a position, according to Bigo, to delude the public into believing that "they are deciding what kind of dominating powers are acceptable or not"[31] even though the suspension of oversight means that the public is no longer in an informed enough position to do much more than accept in blind faith the assertions of those who exercise authority beneath the secrecy-veil of "exceptionalism" and who thus have largely unchecked authority to assert the need for exceptionalism itself to continue.

A compelling illustration of the staged transparency that accompanies what Bigo has described as the "exceptionalism of power" coincided with the renewed interest in the Stanford Prison Experiment shown by social scientists and contemporary artists. Almost directly concurrent with the projects by Reicher and Haslam and by Zmijewski, the U.S. government commissioned an official panel to look into the abuses at Abu Ghraib. The task of assembling the panel fell to Secretary of Defense Donald Rumsfeld even though he arguably had a clear conflict of interest in commissioning the panel. Rumsfeld immediately appointed former secretary of defense James Schlesinger as its chair. Schlesinger was hardly an outsider. In addition to being a former secretary of defense, Schlesinger was also a former director of the CIA, and at the time of his appointment as chair of the Independent Panel to Review DOD Detention Operations he was already serving as a member of President Bush's Homeland Security Advisory Panel and as a consultant to the Department of Defense. Despite his close ties to the Bush administration, his work as the chair of the Independent Panel ultimately had all the appearances of a blunt critical assessment of the Department of Defense, and yet that too is precisely the point.

In 2004 (the same year that *Guantanamo: Honor Bound to Defend Freedom* premiered), Schlesinger not only cited the Stanford Prison Experiment in the *Final Report of the Independent Panel to Review DOD Detention Operations* but also devoted an entire appendix of the report to an exploration of the conceptual parallels between the brutality that erupted in Stanford Prison Experiment and the gross human rights violations that occurred at Abu Ghraib. At some level, Schlesinger's reference to the Stanford Prison Experiment was a logical choice since the U.S. government was keenly interested in Zimbardo's experiment from its inception. In fact, it was the Office of U.S. Naval Research that originally provided funding for the experiment. Although Schlesinger argued that the Stanford Prison Experiment "provides a cautionary tale for all military detention operations,"[32] the actual subtext of the report was one of managing a scandal rather than fundamentally changing policy in the way, for example, that in the aftermath of the Stanford Prison Experiment the American Psychological Associate Ethics Code and the Belmont Report changed the ethical and legal guidelines for doing research.

At some level, the disaster of the Stanford Prison Experiment marked the beginning of a new ethical and legal order that emerged from its dark shadow. But rather than a beginning, the Schlesinger report was by contrast a kind of end. The report was filed, shelved, and forgotten. The lowest-level military personnel were prosecuted, and over a decade later none of the leaders who actually created the environment where human rights abuses were possible have been held accountable. Like Zimbardo, they have simply advanced in their careers or moved into comfortable retirement. And their accomplishment? They set a precedent of allowing human rights violations to remain unchecked and their perpetrators to remain unscathed and shielded by the exceptionalism of power. Let there be no mistake here: we have less to fear from a Donald Rumsfeld than we do from those who will follow in his footsteps, after having carefully studied the precedent he set.

As an operative element of the ban-opticon, such political elusiveness—such self-perpetuating exceptionalism of power—represents a form of the surveillance *dispositif* that is radically different from that conceptualized by Foucault in the theoretical models that he derived from his study of Bentham's panopticon. Moreover, this *dispositif* suggests the need for a rethinking of the conclusions that Zimbardo drew from the Stanford Prison Experiment. If the main ethical critique of the experiment has focused on the creation of a context and environment where abuse is facilitated and where equivocal leadership subtly encouraged participants to adopt abusive and sadistic behavior and led them to commit heinous acts they otherwise would not commit, the idea that guards in the Stanford experiment or at sites like Abu Ghraib and Guantánamo Bay somehow "naturally" inclined toward evil seems woefully inadequate. Rather than suggesting as Zimbardo does that abuse somehow results naturally from the absence of clear guidelines, I would suggest that what the would-be guards in the Stanford Prison Experiment and in Zmijewski's *Repetition* ultimately had in common with the real guards at sites like Abu Ghraib and Guantánamo is that they repeatedly played to expectation and were pulled in by the vacuum of an unwritten but hardly ambiguous script—understanding full well what the implicit directive was, not only tasking them with "control and surveillance"[33] of prisoners while neglecting to provide them with clear guidelines but also, in the case of Guantánamo and Abu Ghraib, placing them at a distant and remote site where neither the public nor existing legal structures would counterbalance the lack of clear guidelines with effective oversight. While we need to ask how such arrangements might be avoided in the future, they are hardly an anomaly, and it behooves us to ask now how they are woven into the very fabric of surveillance more generally.

It is this absence of oversight that ultimately distinguishes the Stanford Prison Experiment from the subsequent BBC Prison Study. It is what distinguishes Guantánamo or Abu Ghraib from the BBC Prison Study as well. For the BBC Prison Study was a made-for-television project. Not only was it staged and filmed entirely inside the "Elstree Film Studios in north London,"[34] but the fifteen selected participants in the study also knew "that they would be constantly observed by television cameras and that anything they did might be shown on national television."[35] They even wore lapel microphones so as to better record their comments. This oversight led to decidedly different results from the Stanford experiment. Instead of guards becoming increasingly sadistic, the prisoners staged a revolt and successfully encouraged a number of the guards to join them in starting a new community within the studio. This was the kind of result that I would suggest was only possible in a contrived arrangement that was vaguely reminiscent of the more provocative moments that Richard Schechner, for example, attempted to cultivate when in the 1970s he and the Performance Group were experimenting with environmental theater. Indeed, when their revolt was over, the prisoners and the guards from the BBC Prison Study had done little more than agree to break character and subvert a prearranged performative matrix.

If I might be permitted the use of a rather simple pun, what the participants did in the BBC Prison Study was to orchestrate something that they hoped would be their big break. It is difficult to imagine a corresponding event in real life. This is because in a real prison both the prisoners and the guards would know that it would only be a question of time before a second wave of guards would appear to reestablish order, and when they arrived, the first order of business would be to turn off the cameras so that the guards could do whatever it takes to get the job done. In such a situation, real prisoners and sympathetic guards might have a moment of Bakhtinian carnival or festive unrule, but nothing close to a new community. Somehow this seems obvious, and in this respect, the BBC Prison Study also seems so contrived that it is surprising that Reicher and Haslam's peers took it seriously. At the very least, one might have expected that their peers would have lauded the study for having indirectly exposed the extent to which Zimbardo's original experiment at Stanford was always a bit contrived itself. But rather than taking hold of that common thread, critics have simply let Zimbardo and Reicher and Haslam battle it out with competing sociological narratives.

Reicher and Haslam have suggested that results like the prisoner revolt lead to at least two major conclusions. First of all, following comments that Zimbardo made at the end of the 1980s, they note that while some guards in the original Stanford Prison Experiment "exploited their power," others—like the guards in their own study—"sided with the prisoners" and others still "were tough but fair." Reicher and Haslam thus conclude that "such diversity sits uneasily with the notion that role acceptance is simply determined by the situation," as Zimbardo had largely implied.[36] Second, they argue that the decision to televise their study ultimately demonstrated "how extreme behaviours can be restrained by rendering actors visible, and hence accountable, to broader or yet-to-be encountered audiences."[37] Here Zimbardo, who has been quite critical of the BBC Prison Study in a number ways that I want to discuss momentarily, concurred and has argued that Reicher and Haslam had in fact confirmed what he had long maintained in his own work to "reduce prisoner abuses": namely, that there is a profound need for "greater surveillance of guard-prisoner interactions" and that prison abuse and violence "can be eliminated if all parties in a prison setting realize that their behaviour is open for scrutiny and evaluation."[38] In short, Zimbardo is an implicit and enthusiastic supporter of an idealistic image of Bentham's notion of the "great tribunal committee of the world."

Not only do the results of the Stanford Prison Experiment and Zmijewski's *Repetition* tend to confirm the above-mentioned conclusions about the effect of cameras in the BBC Prison Study, but the ban-optic structures of Guantánamo and Abu Ghraib do as well. Yet they do so from the opposite end of the spectrum. Downey notes, for example, that in the original Stanford Prison Experiment, guards became "increasingly ruthless and severe" and subjected "their charges to ever more brutal treatment" at night when "they thought the cameras [documenting the experiment] were off."[39] Similarly, Zimbardo argues that one of the key parallels between the abuses in Stanford Prison Experiment and those at Abu Ghraib is that "in both cases, the worst abuses occurred during the night shift, when the guards felt that the authorities noticed them least."[40] It wasn't just the absence of leadership that led to abuse. It was the perception of a deliberate removal of any apparent checks or balance against the limited authority that the guards in the Stanford experiment and that the guards at Abu Ghraib and Guantánamo had all been granted. This point bears repeating. It wasn't just the wink and the nod from higher-ups. It was the turning of the blind eye that led to abuse. But in practical terms, it was the calculated turning of many eyes blind. It was as much the removal of sites of maximum surveillance to obscure locations like Guantánamo Bay, Basra, Abu Ghraib, and unnamed black sites—far from scrutiny of what Bentham called the "great tribunal committee of the world"—as it was leaders consciously closing their eyes that led to a surveillance *dispositif* that accommodated and cultivated abuse. And therein lies a fundamental paradigm: the creation of a maximum surveillance environment that is structured in such a way that its functioning mechanisms and the authorities behind them circumvent actual scrutiny and oversight. At its most basic conceptual level, this is the rough paradigm of the neoliberal surveillance society.

To some extent this paradigm may seem at odds with the close reading of Bentham's descriptions of the panopticon that I did in the first chapter of this book and in particular to my earlier analysis there of the dual function of the "great tribunal committee of the world," which Bentham conceptualized not only as a mechanism for oversight and checks and balance but as a tool for intimidating the visiting public, who, seeing the wretchedness of the prisoners, would thus presumably be scared into lawfulness out of fear of being condemned to a similar fate. I don't want to lose sight of the need for clear mechanisms of checks and balances or, more importantly, of the extent to which in Bentham's theories of the panopticon the idea of being subjected to surveillance was viewed as a form of wretchedness to be avoided at all costs, but one of the important differences between Bentham's eighteenth-century paradigm of surveillance, on the one hand, and the paradigms of contemporary surveillance regimes, on the other, is that a literal "public tribunal" is no longer necessary as a tool of intimidation and hasn't been for a long time.

We have just enough knowledge about what goes on inside prison walls or detention facilities like Guantánamo that we no longer need to see the wretchedness of prisoners to fear their fate. Popular culture has done much to stoke our imagination of what prison life entails, and fear of the imagined but nonetheless unknown is a powerful tool. In this respect, the ban-opticon is the perfection of an arrangement designed to cultivate fear. This arrangement, according to Caoimhe McAvinchey, is in fact central to the birth of the modern penitentiary. In her book *Theatre and Prison*, for example, McAvinchey notes that the walls of the modern penitentiary "not only kept convicts in, but kept the public out" and that the "opaque physical separation of one world from the other" has proven to be far more effective at "hinting at the harsh, horrific, hidden world they enclosed" if only because the walls let the public's imagination "run wild with hellish possibilities."[41] Those possibilities only increase when the established mechanisms of oversight or of checks and balance lend themselves to circumvention.

Extreme Surveillance and the Emergence of the Profit Motive

In their combined effect what sites like Guantánamo Bay, Basra, and Abu Ghraib offer is a disturbing model of surveillance that is significantly more complex than the surprisingly one-dimensional notion of surveillance that Reicher and Haslam used to downplay the relevance of the pervasive presence of cameras in their own study. Nowhere is that one-dimensionality more evident than in Reicher and Haslam's defense of their decision to allow the BBC to broadcast their experiment. "Although it may be unusual to be in a position where anything one does might be broadcast into millions of homes," Reicher and Haslam argue, the BBC Prison Study was in their opinion merely "an extreme example of something that is an increasingly common feature of our everyday lives—namely, surveillance."[42] Indeed, "to those who criticize the BBC study on the grounds that it was broadcast," Reicher and Haslam "respond that surveillance is everywhere (anyway)."[43] But the problems with this line of argument are multiple. First of all, it conceptualizes surveillance exclusively in terms of degrees (e.g., "an extreme example") rather than in terms of type, form, and diversity. While a prison might justifiably be called a maximum surveillance environment and while it thus might exemplify an extreme form of surveillance where specific behaviors are monitored, regulated, and restricted, much of that extreme has to do with the fact that as an institution a prison is a site that is specifically designed to be distinct and separate from society more generally. Its surveillance methods—ostensibly at least—are unique to it as a correctional facility focused on monitoring those who have been prosecuted for criminal behavior.

Reicher and Haslam would certainly not be the first to suggest that the uniqueness of the prison surveillance paradigm no longer exists—if in fact that is what they intended to suggest with their rather sweeping statement that "surveillance is everywhere." Critics like Simone Brown and Gary Marx, for example, have expressed grave concern that prison surveillance has become the paradigmatic norm for surveillance in society more generally and has thus resulted in a blanket criminalization of broad swaths of society regardless of whether the activities of those affected merit intense and intrusive scrutiny and policing. Echoing arguments first formulated by Marx, Brown has recently argued, for example, that "the maximum-security society is a way to conceptualize how the surveillance that was once figured as contained inside the military base or the maximum-security prison ('perimeter security, thick walls with guard towers, spotlights, and a high degree of electronic surveillance') now extends out to the whole of society."[44] But what Brown via Marx attempts here is to lay the groundwork for a fundamental critique of this kind of expansion of the technologies and techniques of prison surveillance across society especially since it disproportionately targets minority communities. Reicher and Haslam, on the other hand, simply use it as a way to justify their own actions and to justify their misleading suggestion that surveillance is just another version of something akin to reality television programming. At one level, the problem with this justification is that it uncritically equates the mechanisms of surveillance employed by correctional facilities with the use of cameras by the television broadcast industry. But at another level, there is something to that equation that tells us a lot about surveillance—even if it does so in a manner that runs counter to what Reicher and Haslam intended.

While surveillance may be everywhere, the multiple types of surveillance that exist in society are a constant reminder that everywhere is not the same. Indeed, prisons themselves—even as examples of a maximum surveillance environment—are not all the same, and neither are their prisoners. At the most basic level, this simple fact explains why the detainees at Camp X-Ray were imprisoned on the island of Guantánamo rather than, for example, at the military's Pre-Trial Confinement Facilities in North Chicago, Illinois, or in New London, Connecticut, or at any number of the military's permanent detention facilities in Georgia, California, Mississippi, Texas, or the District of Columbia. Placing the Guantánamo detainees at any of those sites would fundamentally alter their legal status as well as the oversight requirements that are required by law for state and federal facilities: requirements that, among other things, dictate oversight and are designed to protect prisoners and detainees; dictates that thus shape the forms that surveillance takes at any given facility. But digging beneath this basic fact, I want to suggest that where a project like Reicher and Haslam's BBC Prison Study and "a project" like Guantánamo's Camp X-Ray converge is in the startling ties that both of them have to private enterprise. Indeed, the importance of juxtaposing these two "projects" is that doing so helps to tease out the ways that such ties remain a constant across the spectrum of the surveillance industrial complex. Perhaps those ties are not so surprising in a project like Reicher and Haslam's, whose very title acknowledges its debt to the BBC. But the surprising fact that ties to private enterprise constitute a crucial component of the ban-optic structures of a remote site like Camp X-Ray at Guantánamo tells us volumes about the logic of the ban-opticon under the policies of neoliberalism.

Those policies do a lot to explain the rather awkward lengths to which Reicher and Haslam have to go to justify their research project's otherwise inexplicable ties to the BBC. As any British academic working today knows, the need to seek funding sources outside of the academy is a given and is the direct result of neoliberal government policies that outsource as many state functions to private enterprise as possible and structure the remaining ones in ways that make them beholden to strategies that benefit the business sector. In many respects, the BBC Prison Study is a product of this environment even as Reicher and Haslam try to dance around it. "The project," they insist "can be described as 'original science filmed,'" and they maintain that it was "distinct from 'reality television' enterprises" because they devised the study and proposed it to the BBC rather than the BBC or any other "television company" coming to them, having devised a scenario "with issues of entertainment in mind" and choosing then to invite "academics to comment on them."[45] The problem is that this bit of coded logic begs the question of whether the BBC evaluated Reicher and Haslam's proposal "with issues of entertainment in mind" before agreeing to sign on to the project, or, conversely, of whether Reicher and Haslam, given the need to seek funding from the private sector, tailored their proposed project in its conceptual stages to make it appealing to the BBC. Indeed, this problem ultimately led Zimbardo not only to label the BBC Prison Study a "pseudoexperiment"[46] but also to criticize Reicher and Haslam for conflating "scientific research and its values of objectivity and adherence to the cannons of scientific method with those values of public media that centre on entertainment and profit."[47] In fairness to Reicher and Haslam, Zimbardo is arguably holding them accountable for a structural shift in research funding at British universities that is not of their own making. Focusing on a particular instance of its effects on the academy, Zimbardo inadvertently offers us a critique of the compromising effects of neoliberalism and its naive "one size fits all" idealization of the logic of capitalism.

Those same effects are at play in contemporary surveillance. As I will be arguing at length in the closing chapter of this book: neoliberalism defines contemporary surveillance just as much as surveillance defines neoliberalism. Both are now unimaginable without the other and are intertwined in a self-perpetuating dialectic where neoliberalism sustains surveillance and vice-versa. But focusing for the moment specifically on the effect that neoliberal policies have on surveillance complexes like prisons, we begin to get a much clearer picture of the rationale of the island prison both as a literal geographical structure and also as a conceptual metaphor. For if the logic of neoliberalism is that of reduced regulation and oversight of the private sector and that of a privatization of government services, the island prison is emblematic of what happens when those policies begin to take hold in correctional and security services.

Such effects are widespread in implication for surveillance and for democracy more generally. If as a geographical site of extreme surveillance the island prison is designed to escape oversight and scrutiny, the privatization of the key services within the island prison only escalates the loss of the checks and balances that are vital to a functioning democratic society. The BBC Prison Study may have compromised its scientific "objectivity" by catering to private funding sources. But that is a small consequence compared to what goes on at sites like Guantánamo, where something far more nefarious and calculated was built into the reliance on private enterprise. There the reliance was on private contractors who were not bound by the military code of conduct rules, legal structures that regulate the terms of military engagement, or its protocols for handing prisoners and who were hence able act outside of legal boundaries with impunity. In this respect, the question of whether the involvement of private enterprise and the profit motive lead to dubious compromises is hardly limited to issues of scientific research and method. It segues directly into the most basic issues of democratic justice, human rights, and moral probity.

We might simply suggest that in matters of national security, of policing, of surveillance, and of correctional incarceration, the principles of democracy demand the avoidance of such compromises and require a strict separation between the public and private sectors, that is, between the performative logic of the state and the performative logic of private enterprise. But it is precisely in these areas where such compromises have been a part of Western democracy—or of what we might call "democracy's double"—at least as far back as the end of Dwight Eisenhower's presidency in 1961 when he warned of the dangers of the emerging military industrial complex in his final address to the American people.[48] Many critics might mark that warning as a kind of critique of neoliberalism *avant la lettre*. Tony Perucci, for one, has argued that "the performance of the neoliberal economy . . . depends upon the 'performance complexes' that also produce it: the military-industrial and prison-industrial complexes."[49] What is noteworthy in Perucci's argument is his reference to the two complexes of the neoliberal economy that, not coincidentally, converge on the island of Guantánamo and that there offer us a glimpse not only of the logic governing the structures the military industrial complex about which Eisenhower warned the American people, but also of the prison industrial complex that, in its edging toward increasing privatization, represents what we might justifiably identify as the single most brutal and destructive form of surveillance in Western society.

That form so disproportionately targets minority communities that critics like Michelle Alexander have rightly identified the prison industrial complex as the central enforcement mechanism of what she provocatively describes as "the new Jim Crow"—which I cite here as shorthand for what I would describe as the institutional foundation of a vast ban-optic surveillance complex calculated to subject minorities to excessive monitoring and thus to exclude them from social services, rights, and privileges enjoyed by other members of society. The for-profit prison industry banks on those exclusions and on the social sorting, separation, and isolation—indeed, the institutional segregation—that they themselves orchestrate and sustain. With respect to the prison industrial complex, I would argue, the island prison is thus an apt metaphor for the functioning logic of the private prison industry on the mainland: that grotesque offspring of neoliberal politics whose increasingly ban-optic circumvention of democratic oversight is spearheaded by companies like the Corrections Corporation of America (CCA) and the GEO Group in the United States and G4S and the Serco Group in the United Kingdom.

Looking, then, at the underlying structures and the functioning logic of the model of the ban-opticon that emerges from Guantánamo is ultimately an important preface to a wider argument by analogy in which I want to posit the private prison industry as a paradigmatic example of how surveillance performs in the neoliberal society. With that goal in mind, it behooves us to expound upon Zimbardo's critique of the BBC Prison Study and in particular upon his sense that the scientific credibility of the study was compromised because credible oversight fell prey to the profit motive of private industry. If we extend Zimbardo's sensibility for credible oversight to Guantánamo or Abu Ghraib, which, despite having had the opportunity, Zimbardo never did, we start to uncover how Guantánamo is configured as a paradigmatic example of a post-democratic, ban-optic model of surveillance. Here Zimbardo's own reflections on Abu Ghraib are instructive.

In drawing parallels between the Stanford Prison Experiment and the events at Abu Ghraib, Zimbardo has consistently likened the guards in the Stanford experiment to the MPs at Abu Ghraib and Basra. Yet as reprehensible and as disgusting as the infamous acts of soldiers like Chip Frederick, Charles Graner, and Lynndie England might have been, these enlisted personnel were far from the key figures involved in the worst torture and abuse that was suggested by the leaked photos of them abusing detainees and prisoners. Much of that torture and abuse occurred as prisoners moved into the hands of shadowy private security personnel from private companies like Titan and CACI[50] that were contracted by the U.S. government and had a financial stake not only in procuring actionable intelligence but also in perpetuating a cultivated need for their services. What Zimbardo overlooks is the fact that "private contractors were involved in some of the most serious abuses at Abu Ghraib."[51] The point here is this: the use of private contractors at Abu Ghraib and Guantánamo, especially for purposes of interrogation and intelligence gathering, constitutes perhaps the clearest example of the strategies of what Bigo calls the "exceptionalism of power" to be found in the ban-optic structures of surveillance in force at sites like Guantánamo and Abu Ghraib.

At all of these sites, contractors were able to function without oversight, constraint, or, most importantly, accountability. In short, they had "exceptional" status. Even the military concedes as much in the Fay-Jones report, which, after the abuses at Abu Ghraib became public, Lieutenant General Ricardo Sanchez commissioned to investigate "the Intelligence Activities at Abu Ghraib." In that report, Major General George Fay wrote:

> No doctrine exists to guide interrogators and their intelligence leaders . . . in the contract management or command and control of contractors in a wartime environment. These interrogators and leaders faced numerous issues involving contract management: roles and responsibilities of JIDC [Joint Interrogation and Debriefing Center] personnel with respect to contractors; roles, relationships, and responsibilities of contract linguists and contract interrogators with military personnel; and the methods of disciplining contractor personnel. All of these need to be addressed in future interrogation and interrogation management training.[52]

Abu Ghraib was hardly an isolated incident with regard to contractors having no clear oversight or accountability. Three years later, the FBI—forced by a Freedom of Information Act filing—disclosed comparable allegations about the use of private contractors at Guantánamo, not only placing "private contractors at the center of interrogation operations" but actually indicating that private contractors were "directing the Army's interrogation efforts at the military detention center in Cuba."[53] Reporting for the *Washington Post*, Griff Witte and Renae Merle noted that these disclosures not only "suggest a greater role for contractors than was previously known," they also indicate "a further blurring of the limits on how much responsibility the private sector can carry in doing the public's work."[54]

The consequences of the blurring of these limits can still be felt. As recently as the spring of 2013, for example, Naomi Wolf noted that despite Obama's promise in 2009 "to shut down 'Gitmo,'" the only thing "that the Obama administration has shut down is the office of the special envoy, Daniel Fried, who had been tasked with its closure."[55] In fact, Wolf points out that the 2013 Congress even granted the Pentagon's request for an additional $49 million, not to disassemble Guantánamo, but to build new and additional facilities there instead. Most important of all, however, is the conclusion that Wolf draws. Not only does she suggest that "one reason why the Pentagon needs to build a costly new facility has to do with the role of private contractors in driving detention policy," but in a trenchant and keen criticism she also bridges the island prison that is Guantánamo with the mainland, linking in very short order the profoundly problematic investments that private corporations have in sustaining the current correctional order. Wolf argues that "the vast, often undocumentable profits that flow to these companies [private contractors] go a long way toward explaining why facilities like Gitmo—and privately owned and operated prisons in the US itself—never close. The transfer of public money to private corporations is far more attractive than old-fashioned market capitalism."[56] It has also proven to be more attractive than ethical consideration of the incarcerated men and women whose lives waste away behind bars as the price for those profits. Such is, I would suggest, how the economy of surveillance performs.

Race on *The Island* and Blind Spots in the Prison Experiment

Naomi Wolf is one of the few mainstream journalists to see the line connecting Guantánamo with the private prison industry. But however short the distance from Guantánamo Bay directly to the U.S. coast (536 miles), I want to suggest that the critically richer route veers from the straight and narrow and is a bit more circuitous. So before heading from the military industrial complex at Guantánamo to the prison industrial complex on the U.S. mainland, theater historians and performance studies scholars might plausibly veer toward Robben Island off of the coast of Cape Town, South Africa—that infamous island prison from the apartheid era. For if Guantánamo serves as the backdrop to the early twentieth-first century's most important prison play, that is, Brittain and Slovo's *Guantanamo:*

Honor Bound to Defend Freedom, Robben Island serves as the backdrop to what is arguably the most important prison play of the late twentieth century: Athol Fugard, John Kani, and Winston Ntshona's antiapartheid drama *The Island*. The bridge between these two works of dramatic theater is an important one, particularly since it not only provides direct passage from the island of Guantánamo to Robben Island, and thus links two island prison narratives, but also enters into the sociopolitical territory left uncharted by studies like Zimbardo's Stanford Prison Experiment and Reicher and Haslam's BBC Prison Study. Taken together, the two plays place a number of issues about prisons and surveillance in critical relief. Not the least of these are the blind spots in Zimbardo's prison study regarding the decisive role that assumptions about race and politics arguably play in the processes of dehumanization that are the necessary prerequisite to prisoner abuse and to the most disruptive and suspect forms of surveillance in Western society. Indeed, whatever the insights of Zimbardo's study, the blind spots here loom large over any conclusions one might want to draw from the Stanford Prison Experiment.

The absence of any serious consideration of the role that ideology and politics play in matters of incarceration and prisoner abuse is itself a stunning omission in Zimbardo's original study. Here the question that was never asked in his experiment at Stanford concerns the extent to which social privilege and a cultivated sense of entitlement contributed to the participants' willingness to abuse others. More important still were the broader implications that Zimbardo drew from a social experiment that was conducted in the privileged environment of an elite, private institution to which few minorities then had access. Considering, by contrast, the disproportionate way that the actual prison industrial complex impacts communities of color, the absence of any real consideration of race not just in the experiment itself, but also in real prisoner abuse and in the very governing institutional assumptions of the prison industry as a mode of surveillance is even more striking and problematic. Rather than taking issues of class and race into account, Zimbardo resorted instead to what one might describe as a brand of humanism that accommodated pseudoscientific generalizations about human nature.

Indeed at some level, his conclusions swept across race much in the same way that conservatives recently have answered the call "black lives matter" with the diluted generalization that "all lives matter" so as to avoid ever having to address the unique threats that African Americans face daily from the state-sponsored surveillance regimes exemplified in the police, prison guards, agents of the court, and other representatives of state authority. In this respect, it ought to have come as no surprise that the report James Schlesinger issued on behalf of the Independent Panel to Review DOD Detention Operations devoted so much attention to Zimbardo and the Stanford Prison Experiment. It was far easier to suggest that abuses at sites like Abu Ghraib and Guantánamo were the consequence of human predisposition to conformity and abuse of unchecked power, a so-called "Lucifer Effect," than it was to suggest that a widely cultivated bigotry against Middle Eastern men or people of color more generally was central to the human rights violations perpetrated by the U.S. military.

With such issues of race before us, veering toward *The Island* becomes much more than a detour or digression. The echoes between Robben Island and Guantánamo are as profound as they are neglected—even if they remain exiled, so to speak, from mainstream critical debates. Juxtaposed with a work like *Guantanamo: Honor Bound to Defend Freedom*, Fugard, Kani, and Ntshona's drama, which was written and performed within two years of the original Stanford Prison Experiment, provides a blunt reminder of the role that race plays not just at sites like Guantánamo but in the operative dynamics of the ban-optic structures of the prison industry and of surveillance across society.

That reminder is particularly clear in moments like those when the character Winston tells his fellow prisoner John, "I didn't walk with those men and burn my bloody passbook in front of that police station and have a magistrate send me here for life so that he can dress me up like a woman and make a bloody fool of me,"[57] or earlier when he tells John that their fellow prisoner Sipho is back in solitary confinement because he "complained about the food" and "demanded to see the book of Prison Regulations."[58] As is well known, the plot of *The Island* centers on the two characters, Winston and John, who have been sent to Robben Island for their antiapartheid activism and who, against the backdrop of physically exhausting manual labor, seek to mount a two-person adaptation of Sophocles's *Antigone* (hence Winston's reference to being dressed up "like a woman") that will serve up an allegorical indictment of the prison authorities. If, as Caoimhe McAvinchey has argued, "Thinking about prison helps us understand aspects of theatre and performance,"[59] *The Island*'s play within a play reverses the polarities of McAvinchey's argument and asks us to reflect upon the fundamental theatricality of prison itself and thus, by extension, of the surveillance paradigms it exemplifies. At one level, that theatricality plays out in the moment when Sipho exposes "the book of Prison Regulations" to be a mere prop: a script presented for appearance and show but not as an authoritative source of action. It has little to no effect on the actual food Sipho receives or on the harsh terms of the solitary confinement that the guards dish out to him. Indeed, Sipho's fate is a prime example of the island prison as a site of ban-optic surveillance where "the great tribunal committee of the world" is absented; where extreme modes of surveillance like solitary confinement can be implemented without scrutiny of the duration or conditions of the confinement; and where the book of prison regulations can be referred to for show on the off chance that someone made inquiries. Away from the mainland, the official rules are just so much talk, for there is no oversight or accountability worthy of the name, only the pretense and theater of checks and balances.

If there are echoes between the character Sipho's futile demand to see "the book of Prison Regulations" in *The Island* and the actual "policy vacuum" that Gen. Michael Lehnert rightly worried would (and in fact did) lead to abuses of power at Guantánamo, so too are there eerie echoes between Winston and John's adaptation of Sophocles's *Antigone* and the adaptation of Aeschylus' *Prometheus Bound* that the social theater group Outside the Wire brought to Guantánamo in 2010 in what can only be described as a near surreal gesture at supporting the troops. Outside the Wire, which was cofounded in 2009 by Bryan Doerries and Phyllis Kaufman,[60] retitled their adaptation of Aeschylus's drama *Prometheus in Prison.* The aim in bringing the piece to Guantánamo, Doerries claimed, was "to relate the impact of war on families and relationships, and [thus to] provide a forum to discuss the effects of those experiences, utilizing . . . ancient plays that . . . depict the seen and unseen wounds of battle."[61] Doerries told the military press that Aeschylus's drama is "about how this prisoner rebels against those who have incarcerated him, his family, and friends." The piece thus appeals "to those who serve in the criminal justice field, and especially here at the detention camp," Doerries explained, because its "story focuses on the pressures involved with working with self-righteous prisoners and the impact that has on those involved."[62] But as Doerries was later to concede, the problem is that the critical undercurrent of Aeschylus's narrative runs counter to Doerries's expressed goal. It's hard to support the troops when the gist of the play is pointing a finger of accusation right at them. While *Prometheus Bound* might rightly be described as a "story about power, surveillance, hierarchy, control, and . . . martyrdom,"[63] so too, Doerries finally admitted, does Aeschylus's drama give voice to "one of the strongest condemnations of authoritarian rule, unjust imprisonment, and cruel and unusual punishment in all [of] Western literature."[64] In this respect, beneath the constraints of Doerries's adaptation, Aeschylus's bound Prometheus gives voice to a powerful indictment against the cruel and unusual punishments dished out at Guantánamo: punishments like the forced feeding of prisoners that Doerries actually witnessed[65] and that the UN Human Rights Commission and the World Medical Association have denounced as a form of torture; punishments like the excessive use of solitary confinement, which the U.S. military has attempted to hide from the public with the use of euphemistic descriptions like providing Guantánamo prisoners with greater "privacy" or with "single-occupancy cells."[66]

But what of Winston's reference in *The Island* to the burning of his passbook? While Sipho's stint in solitary finds clear echoes in the ban-optic structures of Guantánamo's Camp X-Ray, Winston's reference to his passbook extends the ban-optic out into society more generally. As Errol Durbach noted back in 1984, "The pass-law system . . . [gave] the State the right to control the movement of Blacks in South Africa" as well as to "compel them to work in a specific area under penalty of being deported to one of the Bantustans," that is, to one of the segregated "homelands" that were designed to enforce apartheid by restricting blacks to specific areas and banning them from others unless they submitted to a rigorous, racially charged surveillance regime, of which the passbook was the cornerstone. All of this, Durbach argues, had "the effect of enslaving the worker to his permit, destroying families, and depriving the Black man of mobility, self-determination, and freedom."[67] We might delude ourselves into believing that the apartheid surveillance regime was some sort of an anomaly from the not so distant past. But it was merely a crudely formulated example of the racist underpinnings that continue to shape surveillance in Western society. In his book *Identifying Citizens: ID Cards as Surveillance,* David Lyon reminds us, for example, that "identification is the starting point of surveillance"[68] and that ID cards, passbooks or whatever one might call them always constitute a central component of "systems for social sorting" (particularly when they are linked with databases), which are designed to permit "extensive discrimination between different populations through modes of classification that . . . often include ethnicity and religion."[69] With regard to South Africa in particular, Lyon notes not only that the passbook "restricted the mobility and life-chances of black Africans" but also that international corporations like IBM had a vested financial interest in the surveillance technologies that supported the passbook system and hence in political policies that would preserve apartheid rule.

Of course, this was nothing new for IBM. As Lyon points out, during World War II IBM "was recruited to provide the technical infrastructure for genocidal identification." Later in South Africa "IBM, along with the UK firm ICL, also had a role in supplying the passbook computer infrastructure" that gave technical support to South Africa's brutal enforcement of apartheid.[70] There is an obvious echo here between private corporate interests that saw financial gain in preserving apartheid and those that see financial gain in preserving Guantánamo's Camp X-Ray. The echoes between Robben Island and Guantánamo are strong. I would suggest that the echoes are in fact strong enough to reach all the way to the U.S. mainland.

What Color Is Surveillance Anyway?

While some historians may see the thematic connection between one island prison narrative and another, they might also be inclined to question attempts to link a drama from a different political culture that lies forty some years in the past to the present realities of the neoliberal prison industry in the twenty-first-century United States. Perhaps it is, in fact, a long journey from Fugard, Kani, and Ntshona's 1973 South African play back to the North American shores and to the U.S. prison industrial complex—a journey that carries faint symbolic echoes of the Middle Passage. But lest one think that the distance between the two is too great, it is worth recalling what Michelle Alexander has noted in her book *The New Jim Crow*, namely that today "the United States imprisons a larger percentage of its black population than South Africa did at the height of apartheid."[71] This statistic alone ought to provide license for the long passage through time and culture. And, of course, there are other statistics as well. It is now common knowledge that the United States has the highest incarceration rate in the world, holding some "25% of the world's prison population, but only 5% of the world's people."[72] So too is it common knowledge that while blacks, Latinos, and Native Americans only make up about 30% of the U.S. population, they make up over 60% of its prison population.[73]

Correlating such statistics with the legacy of apartheid is one thing. Adam Gopnik has chosen instead to correlate them with the history of slavery in the United States. Reminding us that "there are [now] more black men in the grip of the criminal-justice system—in prison, on probation, or on parole—than were in slavery," he argues that mass incarceration is "the fundamental," indeed the defining, social fact of the U.S. today, just "as slavery was the fundamental fact of 1850."[74] Gopnik's assertion that mass incarceration is *the* defining fact of U.S. culture is not only a way of saying that the United States is fundamentally a surveillance state but also of saying that it is a particular kind of surveillance state, one that targets specific communities and one that uses current or past entanglement with correctional services as a mechanism to nullify and circumvent rights that citizens presume they can take for granted. As Alexander has noted, those released on probation or parole "are subject to regular surveillance and monitoring by the police and may be stopped and searched (with or without their consent) for any reason or no reason at all."[75] With roughly one in three Americans having "some type of criminal record,"[76] the stakes in such entanglements with correctional services are high since they ultimately subordinate not just those on probation or parole but those with a criminal record to a life of excessive scrutiny, monitored disqualifications, barrings, and exclusions that would otherwise be illegal. In short, entanglement with the broad tentacles of the prison industrial complex sets the modern U.S. ban-opticon in motion. In this respect, Gopnik's claim that mass incarceration defines America is likely to sound *more* or *less* bold depending upon the community to which one belongs and upon the extent to which that community has been adversely affected by our culture of excessive incarceration and correctional control. It is not just that the United States has some 2.2 million people behind bars, or that there are almost 5 million others who are currently on probation or parole.[77] It is not even that one in three Americans has a criminal record that subordinates them repeatedly to excessive scrutiny and surveillance by public and private institutions and corporations. It is that their families, friends, and associates—entire communities—are pulled into the scope and into the effects of correctional surveillance regimes by mere proximity and association. And it is perfectly obvious which communities are impacted the most.

The "fundamental fact" to which Gopnik refers not only sweeps across communities of color but weaves a very particular type of surveillance into the fabric of their lives and into the perception that they and others have of their cultural and political identity. Although surveillance per se is not the focal point of Alexander's study, this sweeping effect of mass incarceration and of the systems of surveillance of which it is a part on communities of color is the central argument of her book *The New Jim Crow*, where she argues that rather than relying on race, "we use our criminal justice system to label people of color 'criminals' and then engage in all the practices we supposedly left behind."[78] That label of "criminal" defines a person's subject position within a vast system of surveillance techniques that are designed not to watch, observe, and record but to isolate, contain, and exclude. Today, Alexander argues,

> It is perfectly legal to discriminate against criminals in nearly all the ways that it was once legal to discriminate against African Americans. Once you're labeled a felon, the old forms of discrimination—employment discrimination, housing discrimination, denial of the right to vote, denial of educational opportunity, denial of food stamps and other public benefits, and exclusion from jury service—are suddenly legal.[79]

Alexander rightly characterizes this legal process of disenfranchisement as a sophisticated bit of masquerade that, beneath the guise of the objectivity of the law, smuggles Jim Crow back into circulation. One would be hard pressed to find a better description of how the ban-opticon performs, or of how surveillance as a political mechanism in society always gravitates toward disenfranchisement. One would also be hard pressed to find a more precise model of its calculated social sorting or of its rigorous enforcement of the boundaries that such sorting establishes or of the racial segregation that is its dirty secret. But it is not just that such moments of social sorting coincide with the functional crux of the ban-opticon as such. They mark the boundaries of postdemocracy itself, and I would thus suggest that the new Jim Crow is unfortunately the necessary point of departure for theorizing a postpanoptic notion of surveillance. So the conclusions here are as clear as they are disturbing.

No discussion of surveillance is complete without a clear and frank discussion not only of how the prison industrial complex is a key functioning mechanism of the surveillance society as it is constituted in the United States today, but also of how its institutionalized racist underpinnings are in fact a constituting element of U.S. society as a whole. Keeping this point in mind, I want to pause momentarily in the broad pursuit of this chapter's argument. Much of what I have been arguing throughout this chapter is that as a site of maximum surveillance—as a site that is an audacious and cynical realization of the ban-opticon—Guantánamo's Camp X-Ray offers a twenty-first-century paradigmatic alternative to Foucault's classic embrace of the panopticon as a model of surveillance, and in the closing arguments of this chapter, I will return to this paradigm of the ban-opticon to explain how it serves as a conceptual metaphor for the functioning structures of the equally audacious and cynical private prison industry that is spearheaded by companies like the CCA and the GEO Group and that not only profits financially from—but that also staunchly advocates the continuation of—the policies that Alexander denounces as the new Jim Crow. But before doing so, I want to dovetail those arguments with Simone Brown's radical critique of Foucault in her book *Dark Matters: On the Surveillance of Blackness*.

The motivation for this dovetail has to do with the counterparadigm that Brown offers to Foucault's influential reading of Bentham's panopticon. Early in her book, Brown notes that "surveillance is nothing new to black folks,"[80] for it has defined their American experience since the moments of kidnap in Africa that in turn led to the Middle Passage. Her own book, she notes, thus "stems from a questioning of what would happen if some of the ideas occurring in the emerging field of surveillance studies were put into conversation with the enduring archive of transatlantic slavery and its afterlife."[81] So following Brown's critical intervention, one might draw a line that begins just off the coast of the African continent where she locates the generative moments of modern surveillance, and one might trace that line all the way to the United States and to the racialized culture of mass incarceration that Alexander characterizes as the new Jim Crow. Even at a conceptual level, there is warrant for this coupling and for the highlighting that it roughly achieves of the line I earlier envisioned between Robben Island, Guantánamo, and the U.S. mainland.

Brown is cognizant of the fact that, in looking to Foucault and his reading of Bentham's panopticon, surveillance studies scholars have posited the origins of contemporary surveillance at the birth of the modern penitentiary. But rather than following this lead, Brown takes us a further step back in history to the point in 1785 when on one leg of Bentham's long journey to Russia, where he "would first conceive of the Panopticon," he boarded a "Turkish caïque" in Smyrna "in an attempt to reach Constantinople."[82] In addition to its twenty-four other passengers, Bentham noted in a letter to his father that the ship also carried "18 young Negresses (slaves) under the hatches."[83] Taking note of these women in bondage who were banned from the ship's deck, incarcerated below largely away from sight, and treated as cargo, Brown asks, "How must we grapple with the panopticon, with the knowledge that somewhere within the history of its formation are eighteen 'young Negresses' held 'under the hatches'?"[84] It is at this point that Brown offers a radical revision of the scholarly fixation on Bentham's vision of the modern penitentiary and hence also offers a radical revision of one of the key conceptual pillars of surveillance studies. Against Bentham's famous 1791 detailed, graphic design of the panopticon, Brown juxtaposes the now equally famous 1789 detailed, graphic design of the slave ship *Brookes*, which details how one might transport the maximum number of incarcerated slaves with the least amount of threat and supervision. This, as Brown duly notes, is a design of a mobile maximum surveillance environment.

Inasmuch as the birth of the modern penitentiary serves as a generative moment of surveillance, Brown's provocative juxtaposition constitutes a major historical intervention, and it shifts the genealogical trajectory of surveillance, placing race and commercialized racism as a cornerstone in the construction of what we now call the surveillance society. "Drawing a line through panopticism by way of the slave ship is," Brown argues, a "means of interrupting Foucault's reading of discipline, punishment, and the birth of the prison, because, as Marcus Rediker put it, the slave ship was 'a mobile, seagoing prison at a time when the modern prison had not yet been established on land.'"[85] The reverberations of this model echo through sites like Robben Island and Guantánamo—out to sea as they are, so to speak—and they are a reminder of the central role that race plays in the underlying assumptions that have governed their constructions as sites of maximum surveillance in general and of examples of the ban-opticon in particular. Perhaps in an even more literal way, so too do those reverberations echo though the bowels of U.S. ships like the *Bataan*, the *Peleiu*, and the *Stockham*—all of which, as the British journalist and playwright Richard Norton-Taylor reported in 2008, have been implicated as black-site prison ships where detainees in the war on terror were disappeared and tortured.[86] At least seventeen other ships were implicated in this ban-optic system of mobile seagoing prisons as well. The most prominent among them was the *Ashland*, which, amid disquieting echoes of the east African slave trade from centuries before, was operating off of the coast of the African continent in 2007 and which was reported to have held detainees from Somalia, Kenya, and Ethiopia.[87]

Theater historians are likely to be most familiar with Richard Norton-Taylor not as the *Guardian* journalist who broke the story on the above-mentioned seagoing black-site prisons but as one of the documentary theater playwrights who helped to put London's Tricycle Theatre on the map as a center of creative antiwar activism during the ethical and moral disaster that was the Iraq War. Instrumental in forging what has come to be known as the Tricycle Theatre's "Tribunal Plays," Norton-Taylor's rise to prominence in theater, however, predated the war in Iraq and began with his 1999 documentary theater piece *The Colour of Justice: The Lawrence Inquiry*. That play, which chronicles the London Metropolitan Police's gross—indeed, negligent—mishandling of the 1993 case of an eighteen-year-old black student by the name of Stephen Lawrence, who was murdered at a bus stop by white racist youths in southeast London, has an all too familiar ring for anyone possessing a critical understanding of the racial politics of policing in the United States and in the United

Kingdom. As Janelle Reinelt recounts in her article "Toward a Poetics of Theatre and Public Events: The Case of Stephen Lawrence," Lawrence, after being stabbed, "bled to death on the pavement" in part because the police who arrived on the scene "did not give him first aid before the ambulance arrived." In fact, they denied that the unprovoked murder was even a "racist killing."[88] The problem, Reinelt argues, is that "the metropolitan police reacted by treating the incident as if it fit the mold of racial stereotypes instead of its reverse."[89] In short, the police were acting as agents of a system predisposed to see Lawrence as the potential threat rather than as the probable victim.

Today there is no real dispute about the racist underpinnings of Lawrence's murder. But what we are left with is a prime example of the ban-optic at work at the most basic level of street policing, for what Lawrence had every right to expect but did not receive was equal protection under the law by its policing and surveillance apparatuses. What he and his family received instead was a disgraceful illustration of a system whose operative assumption was that he did not belong. Is it really that hard to fit the murder of Stephen Lawrence within the larger frame of the racially charged performances of the ban-opticon? If one looks for the pattern of consistency that links a work like Norton-Taylor's *The Colour of Justice* to his subsequent reporting in articles like "US Accused of Holding Terror Suspects on Prison Ships," one is also likely to find the color of surveillance.

Returning again to Brown's study and to that historical moment when "under the hatches" of the ship that Bentham took to Constantinople eighteen African women were incarcerated and transported as chattel, I want to suggest that the importance of Brown's intervention in the conceptual genealogy of surveillance lies in the fact that she grounds her arguments in a materialist history of race rather than in an abstract concept of it. For the women "under the hatches" were separated not merely because of a cultural antipathy toward blackness as such but also because of a racist economy that profited from their bondage and from classifying them not even as second-class citizens but as private property to be bought, sold, and exploited for financial gain. In practical terms what Brown's focus does is to link racism and private enterprise and thus to place the use of incarceration and extreme modes of surveillance for exploitation and financial gain at the defining moment of modern surveillance as a sociopolitical phenomenon. But for theater historians and performance studies scholars there is also something slightly familiar in the narrative that Brown retells from Bentham's journals. If, as I noted earlier, Foucault's reading of Bentham is one that characterizes the panopticon as a performance machine because it casts prisoners into individual cells that are like "*so many small theatres*, in which each actor is alone, perfectly individualized and constantly visible,"[90] Brown's reading of Bentham's journey to Constantinople is one that implicitly captures the theatricality of the ban-opticon and that with Bentham and the passengers on the deck above and the eighteen slaves trapped below hints at theater's hidden labor. Indeed, Bentham's reference to that which occurs "under the hatch" is vaguely reminiscent of a rare acknowledgment of the stage "trap," that is, the trapdoor beneath which the unacknowledged labor occurs that makes the magic and economy of theater possible.

The labor of stagehands beneath and behind the stage is obviously not coerced in the way that slave labor was. But there is a performance paradigm here through which we might better conceptualize the operative structures of ban-optic modes of surveillance more generally. For just as the great theatrical illusions on the stage are unimaginable without theater's invisible workforce, the social economy through which Bentham moves on the deck of that ship is premised upon the labor obscured and hidden "beneath the hatch." These are the performances that at the most basic level are banned from sight, and, exposed here, they give the lie to the freedom that Bentham enjoys with the other passengers up above. For the difference in the moment recounted from Bentham's journal is not between those who are free and those who are not. It is a difference between the freedom that Bentham believes he enjoys and the theatrical bit of pretense that his freedom actually is.

That freedom is an illusion consumed and enjoyed above only because it is produced by the sweat of those condemned to labor in obscurity below—those who, for the purposes of our social dramas, we pretend do not exist. In this respect, the ship upon which Bentham traveled—that "mobile, seagoing prison"—is in fact a microcosm of the functioning ban-optic structures of modern surveillance economies. The political imperatives here, I would suggest, are clear. It falls to us to dispel the pretense, to open the stage trap from above and to acknowledge the labor below and to acknowledge also the extent to which the security that surveillance ostensibly provides is always a dubious commercial transaction. It is something that is purchased at the expense of those who are subjected to such extreme forms of surveillance that, ironically, they become invisible to the rest of society.

Your Own Private Island: Maximum Surveillance, Private Prisons, and Turning a Blind Eye on Oversight

Brown's focus on the material conditions of those eighteen African women "under the hatches" presents us with a very different picture of how surveillance performs than is suggested by Foucault's reading of Bentham and the panopticon. The model laid out by Brown conceptualizes surveillance as a mode of economic performance and as the governing structure of a market that literally is not free and was never intended to be so, for it thrives on the coercive exploitation of labor, the justification for which coincides with a process of social sorting along racial and ethnic lines. In this respect, it is not difficult conceptually to find the path leading from the moment of slavery that Brown recounts from the late eighteenth century to the contemporary realities of the prison industrial complex that Alexander describes as the new Jim Crow—particularly when one remembers that the Thirteenth Amendment to the Constitution, which abolished slavery in 1865, left one clear loophole. It specifically states: "Neither slavery nor involuntary servitude, except as a punishment for crime whereof the party shall have been duly convicted, shall exist within the United States, or any place subject to their jurisdiction." Even today that amendment serves as the cornerstone to the business endeavors of the private prison industry, to its underlying barbarous logic of prisoners-for-profit, and to the ban-optic structures that secure this hidden economy of captive labor. But the voices of opposition to such arrangements came earlier than the Thirteenth Amendment to the U.S. Constitution even if they did not ultimately prevail. Indeed, one answer to Brown's question about how to "grapple with the Panopticon" after having learned of the "eighteen 'young Negresses' held 'under the hatches'"[91] can be found in the reasons that the British Parliament rejected Bentham's proposal for the panopticon when he presented his idea in the early nineteenth century.

Perhaps the most inexplicable omission in Foucault's reading of Bentham is his failure to note that Bentham's model for the panopticon failed to "gain [British] parliamentary approval in 1810" because, as Bentham conceptualized it, the practical implementation of the panopticon resembled a form of slavery. "Its downfall came," as David Lyon notes, "not on account of the inspection principle" that so fascinated Foucault but on account of the fact that "Bentham had advocated running . . . [the panopticon] as a private prison."[92] The problem Bentham encountered was that his opponents in the Parliament emphatically believed that "prison labour was not to be exploited for profit" and that "the state should control punishment, uncontaminated by commercial gain."[93] Much has obviously changed in the British Houses of Parliament and in the halls of the U.S. Congress since Bentham laid out his proposal, and today amid the intoxicating lure of neoliberal ideologies and the rush toward privatization as an all-encompassing solution to the inconveniences of government bureaucracy his proposal would likely receive—and at some level already has received—a more sympathetic hearing. All but forgotten is the fear that Bentham's contemporaries had that the profit motive would radically alter the priorities and objectives of the correctional process.

The results of that hearing have only confirmed the underlying fears that kept Bentham's contemporaries in Parliament from approving his original proposal. Even today, it is difficult to understand the moral, ethical, or even legal rationale for a system that allows the punishment of wrongdoers to be a viewed by businesses as an opportunity to undercut the wages of law-abiding laborers. But companies like "IBM, Motorola, Compaq, Texas Industries, Honeywell, Microsoft, Boeing, Starbucks, Victoria's Secret, Revlon, and Pierre Cardin"[94] have all taken advantage of the cheap labor that prisons offer at the prisoner's expense. Indeed, as Earl Smith and Angela Hattery have noted, unbeknownst to most citizens, "On any given day the average American uses 30 products that were produced, packaged, or sold out of a prison"[95] for substandard wages. It is one thing to demand that prisoners use their time in prison constructively, but how and where does one find the justification for the use of inadequately compensated captive labor as a means of producing less expensive wares and thus of subsidizing the standard of living of other citizens? Here the disciplined body of the inmate in a private prison is reduced to little more than what Mechthild Nagel has called "neochattel"[96]—a term that is wholly appropriate given the disproportionate ways that communities of color have been affected by such institutionalized coercive exploitation.

Rather than merely citing this kind of arrangement as a nightmarish anomaly, we ought to consider the extent to which the plight of those bound in captive labor—the plight of those subjects of maximum surveillance—is in fact indicative of the underlying economic logic of surveillance across society. Here, once again, we return to the stage trap and to a reminder that the theater—that marvelous social mechanism for watching and being watched—only functions from the labor that is carefully managed and deliberately removed from sight. So if all of the world is a stage—which is as much surveillance's proposition as it is Shakespeare's—it behooves us to look beneath the trap and under the hatches, for there the sight of the theater is not so pretty. Inasmuch as the theater—that site of watching and being watched—is a paradigm of surveillance, so too is it a reminder that surveillance always creates its own calculated and carefully managed shadows, and if we are to truly understand surveillance, we have to dwell less on the stage than beneath its traps, where the economy of the ban-opticon runs the show.

There is, in fact, something epic enough to that economy that one half expects to find Jonathan Jeremiah Peachum manning the trap. In their study of human rights violations and the prison industrial complex, Smith and Hattery have noted, for example, that when prisons are privatized, basic economic logic indicates that the priorities of running a successful and profitable business will inevitably eclipse the correctional imperatives of state-sponsored punishment. "The most basic economic analysis," they note, shows that the private prison "loses money when there are empty cells."[97] So in practical terms, this means that rather than being at the tail end of the judicial process—which begins with legislation, followed by policing, then by prosecution in the courts, and ends finally with prison sentences—correctional services companies like the CCA and the GEO Group have jumped the line, not only providing prison facilities but also lobbying for (and contributing to the campaigns of politicians who favor) legislation that requires stricter and longer sentencing so as to keep the cells in their facilities full. This is no small undertaking. The "CCA has spent an average of nearly $1.4 million per year since

1999 on in-house lobbying, as well as [on lobbying] through high-profile . . . firms such as Akin Gump Strauss Hauer & Feld LLP and McBee Strategic Consulting"[98]—all of it aimed at keeping the billion-dollar private prison industry afloat with legislation that will keep the prisoners coming.[99] In fact, in its 2014 annual report, the CCA told its shareholders: "The demand for our facilities and services could be adversely affected by the relaxation of enforcement efforts, leniency in conviction or parole standards and sentencing practices or through the decriminalization of certain activities that are currently proscribed by our criminal laws."[100] Even more to the point, the CCA "stated in a regulatory filing that keeping the drug war alive is essential to its success as a business: '[A]ny changes with respect to drugs and controlled substances or illegal immigration could affect the number of persons arrested, convicted, and sentenced, thereby potentially reducing demand for correctional facilities to house them.'"[101] This kind of language, which can consistently be found in filings and CCA reports to their shareholders over the past decade, has understandably led Adam Gopnik to conclude that "Brecht could hardly have imagined such a document: a capitalist enterprise that feeds on the misery of man trying as hard as it can to be sure that nothing is done to decrease that misery."[102] Indeed, as a paradigm of modern surveillance we might legitimately label the private prison industry *The Three Penny Correctional Opera(tion).*

Looking beneath the trap of that production, not only are we likely to discover the ways in which the operative logic of surveillance is consistently indistinguishable from that of private enterprise, but we are also likely to see the ways in which the blurring of the two effectively structures the ban-opticon as a mechanism for skirting democratic oversight and transparency. In this respect, it is important to remember that the private prison industry is not autogenous. It is a product of neoliberalism's penchant for deregulation and divestment, a penchant that is itself the product of the intense ideological warfare that for over four decades the business sector has waged against the egalitarian principles of democratic governance. And here we can finally draw the concluding line connecting Guantánamo with the U.S. mainland. For private prisons were born of the same insular logic that later became the raison d'être of sites like Camp X-Ray on Guantánamo.

The inaccessibility of the island prison at Guantánamo, which was a conscious strategy for circumventing proper oversight, replicated what was already the governing structure of the private prison industry as a product of neoliberalism's penchant for deregulation and divestment. Similar connections have been made before. Writing about the gross human rights violations at Abu Ghraib, Tony Perucci, for example, has noted the disturbing ways that "the Abu Ghraib performances . . . can be seen as restagings of the 'training' video produced by guards at the *privately run* Brazoria County Detention Center in Texas, where black inmates were chased and attacked by growling dogs."[103] But that training video, which is itself a restaging of Jim Crow's long history of correctional officers attacking black bodies for sport, is a rare glimpse into the world of private prisons. For what sets private prisons apart from state-run facilities more than anything else is not their supposed cost-effectiveness (which is a myth), but rather the legal mechanisms that exempt them from oversight: "Whereas public departments of corrections on the state and federal levels are subject to disclosure statutes under the Freedom of Information Act (FOIA), private prison firms contracting with public agencies are not."[104] Indeed, the CCA alone has spent over $7 million lobbying against any legislation that would subject private prisons to the standard scrutiny and oversight required at state-run facilities.[105] The consequences here are clear. Since private prisons are not "subject to stringent public records requirements," cases of abuse, negligence, fraud, or racism are difficult to document or counter, for "it can take years to uncover records about recidivism, inmate deaths and suicides."[106] As a result, the private prison industry represents the near perfection of the ban-optic model of surveillance, providing maximum surveillance while deflecting any public oversight worthy of the name.

In that deflection, there is moment of clarity about how surveillance performs and about the contradictions upon which its performances rely. In this respect, the conceptual shift from a panoptic to a ban-optic model of surveillance is about catching those contradictions in motion and play. Foremost among those contradictions are the shadows that surveillance always produces: the unobservable spaces it must create in order to maximize its operative potential. Unfortunately, those are also the spaces where its most egregious acts occur. We can describe those spaces as "under the hatch"—if only never to lose sight of the processes of social sorting that relegate specific communities to surveillance's shadows. There one can only imagine the abuses suffered by those eighteen African women cited by Brown, just as today—even as one of Obama's last executive acts as president was to order the end of the federal use of private prisons, a use that President Trump almost immediately reinstated—one can only imagine the abuses suffered by the undocumented immigrants held in indefinite detention centers now run by the private prison industry. Like the status of their immigration, their safety in surveillance's shadows is uncertain.

ELECTRONIC EYES: CHAPTER 4

Outperforming Activism: Surveillance Technologies beyond the Surveillance Camera Players

> What are you looking at?
>
> —Banksy

An Introduction Somewhere between Three and Five Meters above Street Level

In New York City, street theater is so much a part of the cultural landscape that it's almost as easy to overlook as the increasingly ubiquitous surveillance cameras that now decorate every major street. If you spend enough time wandering the streets of Manhattan, sooner or later you are bound to encounter a naked cowboy, a Rev. Billy, some manifestation of the Living Theatre, or any number of other performance artists who come and go. In the late 1990s and early 2000s, passing by Time Square, Washington Square Park, or St. Patrick's Cathedral, or through the subway stations at Union Square, Astor Place, or Fourteenth and Seventh, there was good chance that you might also have encountered a small group of decidedly amateur performers who attracted your attention precisely because their performances seemed to be directed not toward the public but toward building corners and spaces that were somewhere between three to five meters above eye level. Initially, you might have mistaken them for protesters since they often carried makeshift placards, except that those placards frequently had more in common with the placards found between the frames of silent movies than with the slogans that appear on the signs at political demonstrations. If you paused just long enough to get a sense of who they were or what they were doing, there is a good chance that one of the Surveillance Camera Players would have directed your attention toward a mounted CCTV camera like the one in Time Square owned by Earthcam that has been live-streaming images to the Internet of unsuspecting passersby for two decades and that during a two-year period beginning from 2000 to 2002 became the focal point for the Surveillance Camera Players' production of *You Are Being Watched for Your Own Safety*.

In many respects *You Are Being Watched for Your Own Safety* was typical of the Surveillance Camera Players' work. Based upon a simple agitprop conceit, it could be performed by one to seven performers who stood before a CCTV camera and held up boards upon which they had written:

> YOU ARE BEING WATCHED FOR YOUR OWN SAFETY
> WHO IS WATCHING?
> COPS WITH GUNS ARE WATCHING
> WHO WATCHES THE WATCHERS?
> COPS WILL SHOOT PICTURES THE WAY COPS SHOOT GUNS
> NO MORE RACIAL PROFILING
> NO MORE INVASIONS OF PRIVACY[1]

If the boards in this piece moved increasingly toward direct political statements of protest, other more famous pieces like the Surveillance Camera Players' adaptation of George Orwell's *1984* inclined toward the methods of the silent film and provided either mere summary statements of the narrative or key statements of dialogue from the central characters, as when Winston states, "We are the dead" shortly before he and Julia are arrested. There were a number of other adaptations and original pieces in the Surveillance Camera Players' repertoire, and they performed them throughout New York City from 1996 until 2006. Their debut performance in late 1996 was an adaptation of Alfred Jarry's *Ubu Roi*, and their last recorded performance was in 2009 in Montreal, where they performed an adaption of Denis Beaubois's *In the Event of Amnesia the City Will Recall*, which they simply called *Amnesia*—apparently having forgotten the rest of the title. I want to comment on the piece *Amnesia* toward the end of this chapter, but first I want to offer some speculation about the larger context of the Surveillance Camera Players' decision to disband.

Contrary to what one might expect, their breakup is interesting in large part because it was not particularly spectacular. The Surveillance Camera Players did not come to an end in courtroom prosecutions for civil disobedience. They did not end in a moment of financial crisis. Nor was there a messy breakup of its core members amid internal political disputes. They didn't go out with a bang. They petered out. The simple explanation here is that their demise resulted from evolving interests and commitments among key players like Bill Brown[2] and the other founding members. After the breakup Brown continued to give tours that map out CCTV cameras in NYC and was involved in a massive translation project of the work of Guy Debord, and as Leah Lievrouw has noted, "Videos of their performances have been shown in major contemporary art exhibitions and have been posted on YouTube."[3] But even if new interests led to their breakup, one might legitimately question why the steady growth of surveillance cameras and digital surveillance technologies as industries did not solidify the Surveillance Camera Players' sense of purpose and resolve as an activist ensemble, why it did not inspire dramatically new strategies of activist performance, or why it did not rally the public behind their cause.

Hovering over all of these questions is a larger, more general question of why the Surveillance Camera Players disbanded when everything they were challenging was multiplying in a flurry of unprecedented expansion and technological growth. This question is perhaps more interesting than what their activist theater accomplished during the decade in which they were active. One backstory, seldom discussed and buried in the archives of news reports from more than a decade ago, centers on Brown himself, who was apparently arrested on January 8, 2004, and convicted on charges of misdemeanor aggravated harassment for making harassing phones calls to minors. The arrest came about, according to Samuel Bruchey at *Newsday*, after surveillance cameras helped investigators to identify Brown at the Manhattan law firm of DeBevoise & Plimpton, where a number of the calls originated and where, according to Bruchey, investigators said Brown "had occasionally worked as a temporary proof-reader." According to Bruchey's article, authorities subsequently arrested

Brown "outside a temporary employment agency," and they told Bruchey that Brown "later admitted to making the calls." Following up on these events, Bruchey also reported that, while serving as a member of the faculty at the Rhode Island School of Design, Brown had been arrested some "six times in Rhode Island for calling random numbers in hopes that a young girl would answer the phone."[4] The odd part of this story within the larger history of the Surveillance Camera Players is that as demoralizing as it must have been, the Players continued performance until 2006 before ending their collective endeavors. At a practical level, the other members of the group could have simply carried on without Brown. But without discounting the repulsiveness of the reports about Brown, I want to suggest that there were larger forces at play regardless of his own morally and legally suspect behavior on telephone lines. Indeed, the answers to the question of why the Surveillance Camera Players disbanded when everything they were challenging was multiplying were already evident in the underlying assumptions of their founding.

That founding centers around two pivotal documents, the first of which was a short manifesto entitled "Guerilla Programming of Video Surveillance Equipment."[5] Written in 1995 by the activist Michael Carter, this manifesto gave the Surveillance Camera Players their original focus, and, as they themselves acknowledge, their founding was a response to Carter's call for groups to create scenarios and act them "out using surveillance cameras as if they were their own, as if they were producing their own program, and as if the audience consisted of security personnel, police, school principals, residents of upper-class high security neighborhoods, and the producers and salespeople of the security systems themselves."[6] In his call, Carter noted that "some cameras are monitored live, while others record to tape that will probably be viewed only in the event of some crime. . . . For this reason, [he argued,] Guerrilla actions at 24-hour bank machines aren't too productive."[7] This emphasis on cameras that are "monitored live" was echoed in the Surveillance Camera Players' own founding document a year later.

In the "Programming Note for the First Season" of the Surveillance Camera Players, Bill Brown wrote in late 1996 that the Surveillance Camera Players "are not a professional theatre troupe, nor are they producers of or actors in television; they are just a bunch of average Joes and Josephines who appreciate how boring it must be for law enforcement officers to watch the video images constantly being displayed on the closed-circuit television surveillance systems that perpetually monitor our behavior and appearance all over the city."[8] At the time of the Surveillance Camera Players' founding, their self-proclaimed amateurism was noteworthy because it pushed beyond the conventional distinction between amateur and professional theater by emphasizing instead a stark contrast between the average "Joes and Josephines" of the street and a technologically advanced, increasingly privatized, and diffuse police force. Yet while this contrast may have challenged a more conventional understanding of the amateur-professional divide, it was also a sort of historical anomaly. Although the "Programming Note's" contrast may have been somewhat forward looking in its sense of amateur theater, this was not the case in terms of its conceptions of video surveillance. Rather than actually casting a forward gaze, it tended to look in the opposite direction and captured the last glimmers of a quickly disappearing workforce that was wedded to what is often called first-generation video surveillance technology. Contrasting themselves with "law enforcement officers" who watched "video images constantly being displayed" on CCTV screens, the Surveillance Camera Players thus positioned the founding moment of their activism in opposition to a model of surveillance that by 1996 was already on its way out. More important still, by conceptualizing the agents of surveillance as passive (even bored) spectators, the Surveillance Camera Players overlooked the performative dimensions of surveillance technologies themselves.

Those performative dimensions are evident across the entire spectrum of surveillance technologies, even at the lowest end of technological capabilities. Raymond Wacks implicitly makes this same point within a larger discussion of the ways in which CCTV technologies violate privacy irrespective of whether the camera is state of the art or merely an empty artifice. It is not just "the knowledge that I am being observed by . . . a CCTV camera, that violates my privacy," he argues. "What if the camera is a (now widely available) imitation that credibly simulates the action of the genuine article: flashing light, probing lens, menacing swing?"[9] While legal scholars might debate Wacks's conclusion that "the mere presence of a device that appears to be observing and recording my behavior is surely tantamount to the reality of my unease," there is little question that fake cameras simulating "the action of the genuine article" are performative in nature. But they are the lowest-tech version of performance-based surveillance technologies.

Details at 11:00 O'Clock: From Closed-Circuit to Broadcast TV

In all fairness, those performative dimensions may not have been altogether apparent at the time of the Surveillance Camera Players' founding. But whether they recognized the performative dimensions of surveillance technologies—dimensions that I want to discuss at length later in this chapter—the Surveillance Camera Players did become increasingly aware of the limitations of conceptualizing their work solely as a response to people monitoring video surveillance screens. To some extent, this awareness coincided with their becoming increasingly media-savvy as activists. That savviness appears to have had two phases. First of all, by the early 2000s, they had modified their official stance and amended their "founding documents" with the following statement: "Please note that since the group's founding, the Surveillance Camera Players have changed quite a bit. The SCP no longer consider their primary audience to be the police officers and security guards who monitor the surveillance cameras installed in public places. Today, the SCP concentrate on the people who happen to walk by and see one of their performances."[10] This statement coincided with their later assertion: "There is no better way of convincing people that they are being surveilled than by actually showing them the cameras; and there is no better way of convincing people that protests by individuals or small groups can be effective than by encouraging them to join the performance that caught their attention in the first place."[11] It is following this logic that John McGrath has argued that the Surveillance Camera Players' "target audience is largely passers-by, but the performances are humorously constructed for the supposed eye of the security guard watching the monitor."[12] This same logic is behind Elise Morrison's appreciation of calculated effectiveness "in the clarity and simplicity with which the SCP addressed subjects and purveyors of surveillance."[13]

More recently, however, Bill Brown has suggested that the

changes ultimately were geared toward a much larger public audience that could only be reached through a cultivated relationship with the media. He notes that "the SCP didn't simply perform or, rather, it didn't simply perform in front of surveillance cameras and the passersby who might happen to see the performance in action. It also quite deliberately performed in front of the mass media: writers, radio broadcasters, TV crews."[14] Indeed, he argues that "the great power of the SCP—only a handful of people, and yet written about or on TV in a half-dozen languages all over the world—derived in part from the fact that it used to do both types of performances."[15] In short, they performed for those monitoring CCTV screens, for the public that happened to pass by, and for members of the media to whom they not only gave advance notice of their performances but with whom they would also coordinate in order "to accommodate" their own needs "and the needs of the media."[16] This included post-performance interviews with the media about the significance of the performance that they had just seen. By their own account, the group spent "more time explaining and clarifying what it . . . [did] than actually doing it,"[17] and these explanations were formulated with the media in mind. As Brown notes, in their dealings with the media, the Surveillance Camera Players calculated their remarks so that they "were almost always concise, interesting, and 'quotable,'" which ensured that "the media would actually "cover" them, put them "in print" or "on the air."[18]

Inasmuch as the Surveillance Camera Players sought to cultivate broad awareness of the growing presence of CCTV cameras and of the invasion of privacy that the use of those cameras entails, their interest in the media is understandable. This interest clearly paid off in terms of the international audience that they gained for their work and for the issues that their work addresses—an audience that continues to show interest in the Surveillance Camera Players' work well over a decade after they stopped performing. In this respect, they were extremely successful at calling attention to the growing presence of surveillance cameras in public spaces—a presence that occurred largely without prior public debate or referendum. Indeed, debates about surveillance cameras—even the debates provoked by the Surveillance Camera Players—consistently occur after cameras have already been installed. This is, in fact, a consistent narrative for surveillance technologies more generally. Public debate about those technologies tends to be reactionary, following rather than preceding implementation.

Since it is harder to undo what has already been done, or to remove what is already present, the Surveillance Camera Players' interest in the media was clearly an interest in one of the few effective tools left at their disposal for generating public sentiment against an ever-increasing invasion of privacy. To be clear, the decision to play to the media was thus part of a legitimate, calculated strategy for achieving maximum political effect in a struggle that was already weighted against them. That said, a relationship with the media today comes only with compromise and a price since by and large the media is a commercial enterprise, and its primary interests are in sustaining if not increasing its revenues.

One of the things that made the Surveillance Camera Players' work "quotable" is that despite their genuine activist undercurrents, they catered to the kind of mainstream sentiments about surveillance that the media likes. Perhaps there was a bit of double-dipping in this, and at one level critics could argue that the Surveillance Camera Players were able to play to the media's interests while subtly injecting a hard activist agenda into the subtexts of quotable sound bites and images suitable for the evening news. Gabriella Giannachi, for example, has argued that "by feeding art back to the camera, the players literally interrupt the camera's recording of everyday life and thereby sabotage its attempt to transform us from citizens to consumers."[19] At another level, however, the compromises of being "quotable" had their costs. As Morrison has noted, "The SCP tended not to account for the ways in which an individual's gender, race and economic class might affect his or her relationship with surveillance systems and the laborers that 'man' them."[20] Instead, much of the Surveillance Camera Players' work was arguably structured around something out of a populist imaginary: that is, out of an image of surveillance culled from panopticon obsessions and Orwellian allegorical fantasies—both of which are great as provocative metaphors in agitprop theater but neither of which offers a particularly realistic model against which one can structure an effective political activism against the increasingly pervasive technologies of surveillance. Nowhere was this populist imaginary better exemplified than in the Surveillance Camera Players' production of *God's Eyes Here on Earth* and in their adaptation of George Orwell's *1984*.

Crime Deterrence and the Panoptic God

In many respects, *God's Eyes Here on Earth* was the more original and provocative of the two works—even if, at one level, it did play into the Judeo-Christian discursive economy that I have questioned my first chapter. But there is clearly an incisive tongue-and-cheek quality to the Surveillance Camera Players' participation in that economy. A site-specific piece "designed to be performed in front of surveillance cameras used to monitor church property," it juggled irony and analogy in a moment of irreverent protest at a site that many might have considered to be out of bounds as a focus of antisurveillance activism. First performed on July 16, 2000, in front of St. Patrick's Cathedral in New York City, the piece involved one of the players holding up a series of illustrated boards while other members sometimes kneeled as if praying to the surveillance cameras. On the boards the following statements were written:

> Why are there surveillance cameras at the church?
> Doesn't God see everything?
> Daddy says the cameras are God's eyes
> Now I pray to the cameras
> I want God to see me[21]

If nothing else, the breach of conventional decorum that comes with demonstrations that target places of worship was enough to attract the media's attention. If the goal here was simply to provoke debate about the increasing presence of surveillance cameras in democratic societies, then the piece was arguably a success. It certainly was sensational enough to spice up the evening news with images of antisurveillance activism that viewers would likely remember.

So too was the highlighted irony of a church opening its doors to all but obviously trusting none. But in this respect, St. Patrick's was hardly an isolated case. Video surveillance companies market to church organizations all the time. They view them as clients who are every bit as valuable as any other retailer, except that churches apparently face unique security risks since they consort with "publicans and sinners." The company VideoSurveillance.com, for example, specifically argues in its advertisements to churches that "churches can benefit greatly from the security provided by video surveillance. Churches and other places of worship host people of all types and reach out to many groups of people. This open environment can lead to many security issues for churches, including vandalism and theft."[22] Not only does this kind of frank market logic add a whole new layer to the challenge that Jesus issued to moneychangers in the temple when he denounced them by crying out, "My Temple will be called a house of prayer for all nations, but you have turned it into a den of thieves."[23] It also touches upon the "urban entrepreneurialism" that public policy scholars like William Webster, Eric Töpfer, Francisco Klauser. and Charles Raab have argued is behind the emergence of video surveillance or CCTV in public spaces. Indeed, they argue that the emergence of CCTV cameras consistently tends to be directed not toward the public welfare or good but toward potential investors and customers to whom CCTV is marketed as a sign of emerging safety and attractiveness in sites that can then be promoted as viable "places of consumption."[24] I want to talk at length about how the promotion of video surveillance by private enterprise constitutes the single greatest challenge faced by antisurveillance activists and artists like the Surveillance Camera Players, but first it is worth looking beyond the basic debate that *God's Eyes on Earth* encouraged in order to consider how the piece subtly catered to the popular notions of surveillance that mainstream media outlets tend to cultivate and support. The issue here is not the debate about video surveillance per se but how catering to the media framed the debate itself.

Above all, *God's Eyes on Earth* played upon but did not dismantle an analogy that likened the psychological effect of surveillance cameras to the Christian belief that God is basically the ultimate surveillance operator. He is that bored police officer with the long white beard. Not only does God presumably monitor the sinfulness of our actions, but Christians internalize the belief that God is watching them and, in turn, allow that belief to influence their behavior. The Surveillance Camera Players formulated arguments to this very effect in their essay "Time in the Shadows of Anonymity: Against Surveillance, Transparency and Globalized Capitalism." There they argue:

> In *our* society, there is "no darkness"—or, rather, there is "less darkness"—in part because the mass of people have been conditioned to "voluntarily" make themselves transparent to the gaze of all. As

demonstrated by certain works by Michel Foucault and Jean Baudrillard, the Catholic Church (specifically, the institution of confession, which has become generalized throughout "Christian" culture) and "reality-based" television shows, respectively, inculcate and reward the adoption and internalization of the notion that it is both moral and healthy to routinely render oneself transparent (to either the eyes of God or the eyes of the camera).[25]

The similarities between the analogy drawn in *God's Eyes on Earth* and popularly disseminated notions of panopticism are easy to see. Critics like Claire Valier, for example, have specifically linked what she calls the "counter-aesthetics" of the Surveillance Camera Players to "moments which . . . are disruptive of the panoptical."[26] I draw attention to those similarities because they tended to reinforce—or at the very least, did not challenge—the widespread belief among the public that the threat posed by the presence of surveillance cameras (like the threat of eternal damnation from an all-seeing God) will cause people to act within the boundaries of the law. In this respect, *God's Eyes on Earth* played into the kind of quotable populist images of surveillance that unfortunately have dominated much of the public debate about surveillance cameras over the past two decades.

Those debates about the presence of surveillance cameras in public spaces have tended to weigh the invasion of privacy that they obviously represent against the benefits that they presumably provide—benefits that, according to surveillance camera advocates, have direct parallels with the psychological effects that Bentham's panopticon supposedly achieved. While police have embraced surveillance cameras primarily for their ability to record and thus provide evidence for the prosecution of crimes, acceptance of surveillance cameras by the general public is largely the result of arguments that the cameras, because they so readily provide evidence for prosecution, actually play a crucial role in the prevention of crime. Indeed, the surveillance industry has sold acceptance of the cameras to the general public by arguing that they have a panoptic effect and that they thus prevent crime because individuals internalize the awareness of their presence and modify their behavior accordingly. Yet the assertion that the presence of surveillance cameras reduces crime by causing potential criminals to modify their behavior is an argument that scholars have long disputed. Well before the Surveillance Camera

Players staged their first performance of *God's Eyes on Earth*, for example, Clive Norris and Gary Armstrong, in their pathbreaking study of CCTV, cautioned against assumptions that equate the effects of CCTV cameras with what Michel Foucault called "panopticism."

In *The Maximum Surveillance Society* (1999), Norris and Armstrong argue that while "CCTV systems utilize visual surveillance and [often] rely on centralized monitoring"—which are two pivotal aspects of the panopticon—the effects are quite different. Indeed, they question the assumption that CCTV produces the "automatic functioning of power" that Foucault associates with the effects of the panopticon,[27] and that surveillance camera advocates assume when they claim that surveillance cameras function as a crime deterrent. Norris and Armstrong maintain that while "Foucault argues that the strength of the Panopticon design is through inducing anticipatory and, eventually habituated, conformity," the same cannot be said of CCTV. "Unlike the fluctuating and transitory populations of the city centre streets," they argue, "the inmates at total institutions are 'captive' and cannot escape the surveillance gaze."[28] "Because of the certainty of detection, and therefore intervention, is so high inmates may soon learn the futility of resistance and the necessity of conformity."[29] With CCTV, on the other hand, Norris and Armstrong argue that the habituated mental conformity that Foucault speaks of with regard to panopticism cannot be assured because "intervention [by figures of authority like the police] may not be automatic" and is "contingent on a whole host of social processes," such as "whether the screens are being monitored," whether an incident is actually perceived to be "deviant," and whether individuals are recognized as previous offenders and/or potential threats because of behavior or appearance.[30]

Norris and Armstrong were among the very first to offer a serious rebuttal to those who defend the use of surveillance cameras as a meaningful crime deterrent, and their arguments are particularly helpful here because they also highlight the extent to which the Surveillance Camera Players were entangled in debates about surveillance, panopticism, and crime deterrence that continue to have wide popular appeal in the media today and that despite a preponderance of reliable statistical evidence show no sign of abating. That continued appeal, I would suggest, results from the fact that the focus on crime deterrence as a point of debate tends to place the threat of a cultivated mental conformity—and hence the threat of surveillance cameras—at one step removed from average citizens. The logic here is that the panoptic effect on those who view themselves as law-abiding citizens presumably would be minimal since they need not internalize once again a set of values that they believe they already possess—values like those presumably cultivated within church communities like St. Patrick's in New York.

The Surveillance Camera Players' entanglement in these debates played out in the difficult negotiations that individual members of the Surveillance Camera Players like Bill Brown had to conduct between their bourgeoning artistic aspirations and their already established political sensibilities. As Brown has noted, he sees himself not as a "politically inspired artist" but as an "artistically inspired political person,"[31] and if I may, I want to pause long enough to give this statement its due. For as I look back over the process of writing this entire book, it is difficult for me to find a moment that has been more transformative or that has had such a lasting effect on my thinking about art, activism, and the cultures of surveillance than the email exchange in which Brown offered me this point of clarity about his work. Perhaps, the distinction in his email between politically inspired art and artistically inspired activism was a spontaneous, passing reflection, but the distinction Brown made, which he made specifically as a point of clarity about his work and about the work of the Surveillance Camera Players, is in fact a distinction that points more generally toward the required structures of art in the age of surveillance. It points, in short, toward what in the final chapter of this book I characterize as performance postdemocracy.

Coming back to the trajectory of this chapter's argument, one can recognize that the semantic difference that Brown articulated in his clever distinction between politically inspired art and artistically inspired activism is indicative of the larger challenge that the Surveillance Camera Players had in squaring the ambiguities of artistic expression with the clarity of their political convictions. While on the one hand, artistic pieces like *God's Eyes on Earth* tended to reinforce the popularized notions of panopticism that underlie the idea that surveillance cameras deter crime, the actual political statements coming from the Surveillance Camera Players were, on the other hand, very much in line with Norris and Armstrong's original arguments. In published essays like "Deterrence as Crazy Poker Game," for example, Brown and the other members of the Surveillance Camera Players repeatedly challenge the assumption that surveillance cameras deter crime, and they argue that at best the cameras simply "displace crime to places in which potential criminals think there are no cameras in operation."[32] Evidence suggests that they are right. Almost two decades after Norris and Armstrong's pathbreaking study, crime statistics continue—overwhelmingly—to refute claims that surveillance cameras deter crime.[33] But the question here is not whether the cameras deter or displace crime, or even whether the Surveillance Camera Players found their artistic aspirations at odds with their larger political sensibilities.

The question is whether being caught up in the debate about deterrence plays into the kind of fixed populist imaginary that distracts attention from the larger political implications of the increasing pervasiveness of surveillance cameras in Western societies. William Webster argues, for instance, that "crime statistics are 'in many ways a red herring'" for explaining "the impetus behind the revolution or [for identifying] the true beneficiaries of CCTV."[34] If Webster's assessment is correct, that impetus has less to do with deterring crime than with stifling political and cultural dissent—the kind of dissent, in fact, that includes the work that the Surveillance Camera Players did before they disbanded. Examination of CCTV policy, according to Webster, reveals a focus that shifted long ago from crime prevention to the reduction of what is perceived to be "antisocial and undesirable behaviour."[35] In political terms, the implications of that shift, I would suggest, are hard to miss. At some level, crime is simply a liability: an unfortunate but calculable cost of doing business. It's important to reduce that cost where one can, but ultimately the cost can be passed on to the consumer. Antisocial and undesirable behavior, on the other hand—particularly when manifested as political activism, civil disobedience, or cultural dissent—threatens the very stability of the social-economic order as such.

Although those who are inclined toward criminal behavior are likely to pursue it regardless of presumed deterrents like CCTV, the presumption behind the open display of video surveillance cameras is, to follow the logic of Webster's argument, that those who believe in the rule of law but are out of step with the mainstream political order can be influenced or intimidated by the presence of cameras. In this respect, one could do far worse than to ask whether the greater concern behind the installation of video surveillance cameras in factories is pilferage on assembly lines or the intimidation of workers and would-be union organizers. Indeed, such questions not only begin to point toward those whom Webster vaguely identifies as "the true beneficiaries of CCTV," but they also highlight the second major way that the Surveillance Camera Players structured their work around populist notions of surveillance.

Looking for Big Brother's Private Keeper

Inasmuch as the Surveillance Camera Players conceptualized pieces like *God's Eyes on Earth* around popularly disseminated notions of panopticism, so too did works like their adaptation of George Orwell's *1984* play into what is arguably the most established myth about surveillance. That myth has provided untold ideological cover for the private expansion of surveillance technologies because it frames the threat they pose primarily as a threat of state-centered totalitarianism rather than as a threat of technologies controlled by private corporations that are allowed to conduct surveillance without clear and decisive democratic oversight. Indeed, Torin Monahan has argued simply that "the primary adversary for SCP is the state."[36] Ironically, however, Orwell's disturbing tale of obsessive state control—while significant within its original historical context—has subsequently provided cover for antigovernment private enterprises whose combined effect may be even more insidious. One can easily imagine how an adaptation of *1984* might reflect upon the historical transformation in the politics of Orwell's classic narrative. But such reflections were not part of the Surveillance Camera Players' performances of *1984* in the New York City subways.

Given that the pivotal moment in those performances occurred when one of the players held up a board with the caption "Big Brother is Watching You!" the argument by analogy was relatively clear: acceptance of CCTV cameras leads to something akin to the state-centered totalitarian regime depicted in Orwell's dystopian novel. While this populist image plays well to the media—and is arguably the most famous image from the Surveillance Camera Players[37]—it has little correspondence with the realities of surveillance cultures today. If anything, it offers a false sense of complacency because it posits an Orwellian dystopia as a looming but not yet realized threat. But it doesn't take much reflection to realize that the actual threat posed by CCTV cameras results precisely because they are not centralized in the hands of an Orwellian totalitarian state where they might ultimately be subject to oversight and regulation—and it is worth noting in passing that the narrative past tense of Orwell's novel actually implies that Big Brother's totalitarian regime finally succumbs to the rule of law. In short, the novel *1984* is narrated from a "post-1984" perspective. Thirty some years after the actual 1984, however, Orwellian fictions are mostly notable for their flat one-dimensionality. In reality, "Big Brother" is legion.

The real threats from surveillance technologies result because of the largely unregulated—indeed unregulatable—dissemination of surveillance camera technologies into the hands of diverse private commercial and corporate interests. Not only do those commercial and corporate interests have great liberty to collect, evaluate, sort, package, and distribute data without meaningful oversight by the state, but they are able to do so for a number of complex reasons. For example, the highly specialized development of surveillance technologies has occurred at a speed that continues to outpace society's ability to produce legislation that would provide reasonable oversight and regulation of it. But the speed of technological development is only one factor in this equation.

As is often the case with private commercial interests, those interests behind the expansive development of surveillance technologies have considerable lobbying power at their disposal. In no small part, that power results from the fact that those who develop and market surveillance technologies have various state organizations as key clients, and from the fact those clients are, in turn, in a position to provide them with complicated legal protection that, under the guise of "national security," protects their anonymity and privacy. This is especially true in the post-9/11 era, which, as Dana Priest and William Arkin have documented in their book *Top Secret America*, has been marked by an almost incomprehensible extension of U.S. government security designations like "classified," "secret" and "top secret" to private companies and corporations. Among the private companies now benefiting from the newly found anonymity and privacy that such designations confer, companies specializing in surveillance technologies are easy to find—if only because they are so numerous. But what specific services they provide is virtually impossible to determine since those services are covered by the impenetrable screen of state secrecy. Priest and Arkin lay the foundation for a compelling case that this broad extension of secrecy to private companies has radically transformed the fundamental dynamics of secrecy, privacy, and anonymity as concepts. In short, while the Surveillance Camera Players focused much of their activism on issues of privacy, digital surveillance technologies and the companies that develop them were radically redefining anonymity and privacy as such.

It would be hard to overstate the significance of those redefinitions. In many respects, they are key to understanding the politics of new surveillance technologies. In her discussion of the evolution of CCTV, for example, Heather Cameron notes that anonymity itself "is becoming increasingly suspect."[38] In the not so distant past, she argues, "People were more anonymous under the gaze of the cameras. They were being looked at but not tracked." What has changed is that "new technology" is now able to connect "a person's name and database records to their body."[39] The digital software behind face recognition technologies is as capable of generating and maintaining "VIP lists" as it is of "black lists." It is not only capable of recording the faces of those who conduct bank transactions; it also is capable of verifying their identity as the account holder or as the owner of any item that a client wishes to tag and trace. Moreover, the software is capable of constantly updating itself, identifying frequency of use or visits, which in monitoring employees, customers, or patients can then be labeled as sufficient, insufficient, or even as suspect or threatening behavior.[40] But it is not just technologies that have changed. Anonymity itself has changed. As anonymity has decreased with regard to those under the gaze of the digital cameras, so too has it increased with regard to those who own and operate not just the cameras but also the technologies and software that have radically advanced their capabilities.

Beyond the advantages conferred by state designations like "secret" or "top secret," much of that anonymity comes from the fact that private businesses have a financial interest in protecting the substantial investments that they have made in surveillance technologies. They maintain their client base by maintaining ownership of advanced surveillance technologies that they service themselves so as not lose control of the intellectual property of the technology itself or the data that it is capable of gathering and sorting. They lease that technology to various companies and government agencies who enjoy its benefits but do not have ultimate legal control over the technology or in many cases the data itself. In this regard, companies specializing in digital surveillance technologies often remain one or two steps removed from accessibility and control, not only hidden and working behind the scenes of their clients but actively exploring—under the protected auspices of state-sanctioned secrecy—new technological methods for gathering and sorting surveillance data in ways that can be marketed to the ever increasing demands of current and potential commercial clients. As scholars like Cameron have noted, it has been the business community's interest in profiling consumers—what she calls digital "Customer Relationship Management (CRM) techniques"—that has advanced the very technologies that private companies are now selling to governments in order to profile those who seemingly pose a potential political threat. "The data mining and warehousing techniques used to build composite pictures of customers . . . [in order] to sell them more products and services" is now being repackaged and sorted without the customers' knowledge in order to identify "patterns which are claimed to constitute terrorist activity" and can thus be sold to government agencies.[41]

Whether those patterns are the product of scientifically grounded social, political, and cultural research or of gross bias, bigotry, and stereotype is often difficult to say because the surveillance technology providers so frequently remain in the shadows, and it is at this point that we begin to see the new forms of anonymity that emerge as a consequence of privately developed technologies that outpace traditional forms of policing by state organizations. As Cameron notes: "Since the police cannot and will not spend their resources this way CCTV required an outsourcing to private security and monitoring firms who then had an interest in the expansion of their business."[42] As I stated earlier, much of that expansion depends, ironically, upon a vigorous defense of highly limited access to—and hence the privacy of—their own surveillance technologies. In practical terms, this means that a key issue in antisurveillance activism has to be an equally vigorous effort to identify the financial players controlling, marketing, and selling visual surveillance technologies, and this is one of the key areas where the work of the Surveillance Camera Players faltered.

Whereas the Surveillance Camera Players trafficked in images of solitary and lonely "law enforcement officers," mystified surveillance deities, and malevolent older siblings, surveillance studies scholars like Luis Fernandez and Laura Huey have argued that if antisurveillance activism is to remain a viable possibility in the face of increasingly sophisticated surveillance technology and techniques, activists and intellectuals need to shift their attention from techniques and technologies like surveillance cameras and focus instead on enumerating the actual "different types of actors who may engage in surveilling a subject." There is, or course, no uniformity among those actors, and this is the problem with Big Brother as an image. Unlike the Orwellian figure of centralized control, those with the means to surveil, Fernandez and Huey argue, originate from different centers of "power" and "different locations." They can be actors of the state, of public and private institutions, or of corporations. They can even be "individual and interpersonal actors (such as spouses, lovers, neighbours, and so on)."[43]

From a political perspective, there is much to be gained from the shift in focus that Fernandez and Huey advocate, but I would suggest that identifying the multiple actors of surveillance is not altogether separable from knowing what technologies of surveillance are at their disposal or what new technologies they promote—and this brings us back to the Surveillance Camera Players' founding documents and in particular back to Michael Carter's argument in "Guerilla Programming of Video Surveillance Equipment" (1995) that "Guerrilla actions at 24-hour bank machines aren't too productive" because the machines "record to tape." Historically, that argument is important if only because it singled out as "unproductive" actions that targeted precisely the kind of private, moneyed interests and surveillance technologies that were on the cusp of a digital revolution. Indeed, that revolution has literally redefined video surveillance as such. Not only has the advent of digital surveillance technologies eclipsed the populist images of surveillance upon which much of the work of the Surveillance Camera Players relied, but it is also still altering the performative dimensions of everyday life in Western urban society.

Performance, Proprietary Software, and the New Privacy

At the most basic level, this revolution in surveillance technologies—a revolution that accelerated in the aftermath of 9/11 and continues with no foreseeable end—has dramatically multiplied both the actual sites and the principal methods of surveillance. It is one thing for Fernandez and Huey to shift attention to the multiple actors of surveillance. But as they themselves note, it is a mistake to think that one can speak of those actors without also speaking of resistance to surveillance or, I would add, without also speaking of emergent technologies. For neither the continued multiplication of surveillance techniques nor their increasing sophistication occurs within a sociopolitical vacuum. Both are part of a thriving surveillance industry that evolves by catering to the particular needs of individual clients, and those needs frequently emerge in tandem with innovative strategies to resist or circumvent surveillance. Indeed, Fernandez and Huey, like Michael Hardt and Antonio Negri before them, note that resistance to surveillance and the development of new surveillance technologies go hand in hand—the more innovative the resistance, the more innovative the next generation of surveillance technologies and techniques.

We can speak at great length and with legitimate alarm, for example, about how the advent of digital technologies and the algorithms that accompany those technologies have radically increased the sophistication of surveillance technologies like face-recognition software, behavior prediction software, and biometric analytic software. These digital or second-generation technologies are directly traceable back to the kind of technology used by those twenty-four-hour bank machines that Carter mentioned in "Guerilla Programming of Video Surveillance Equipment" and used to keep video records of individual clients and users. But in practical political terms, it is worth considering whether such surveillance technologies and techniques can potentially be read in reverse, that is, in ways that help to identify the agents of surveillance and their motivations, or, more simply, in ways that identify those whom Webster describes as "the true beneficiaries of CCTV." Find out how those technologies perform; find out what kind of information they deliver and how the information is sorted; and, you'll have a pretty good idea of whose interests they ultimately serve. Seldom if ever do they simply serve the individual interests of a flunky police officer monitoring a CCTV screen.

While the advent of such digitally based software generally renders that individual police officer in front of the television screen obsolete, so too does it subject other employees like software managers to an internal surveillance by the very technologies that they maintain. The same technology that monitors perceived threats from the public also has the capability of flagging recommendations that have not been addressed and thus of monitoring the action or inaction of those who are responsible for handling recommendations. Simple though such capabilities may be, they add a profoundly ironic twist to the question of who watches the watchers. For ultimately, those who watch the watchers are the anonymous beneficiaries of surveillance acts: those who are twice removed into autonomous anonymity—at least until they themselves step out into the public arena or have been outed by activists innovative enough to read technology in reverse.

But such moments of countersurveillance activism—such moments of reading technology in reverse—are increasingly difficult, as Lucas Introna and David Wood have noted. They have argued, for example, that one of the most important aspects of CCTV's face recognition software is "the obscurity of its operation." In a very literal sense the software is less accessible than the opera pit in a Wagnerian total work of art. Not only are its algorithms "based on very sophisticated statistical methods that only a handful of experts can interpret and understand," but most face recognition systems also operate with proprietary software—which means that even if someone possesses the expertise to understand them, intellectual property laws restrict access to the software.[44] Those laws touch upon what is arguably one of the deepest contradictions governing surveillance societies today: the software that surreptitiously records and decodes images of individuals without their permission is protected by laws that limit access to it to those to whom the owners have given explicit permission. The software belongs to private rather than public domain. In short, the very software that enables almost incomprehensible invasions of privacy is protected by laws that guard its privacy. This is the new privacy. It is a postdemocracy privacy. In a very literal sense, it is a posthuman privacy as well.

The Performance of Surveillance Technologies

What is of significance to performance studies scholars with regard to these new realms of privacy is that they also facilitate and rely upon modes of performance that are in a perpetual state of development, evolution, and flux. The emergence of these perpetually evolving modes of performance provides an important context for understanding the Surveillance Camera Players' ultimate demise. Ironically, amid their radical understanding of the politics of CCTV surveillance, the Surveillance Camera Players' artistic work remained within a noticeably limited, conventional, and stagnant notion of performance as a concept. Indeed, the Surveillance Camera Players were never far removed from Brown's own description of himself as an "artistically inspired political person" rather than a "politically inspired artist." The distinction here is decisive, for the pivotal question for artist-activists—for politically inspired artists—who are concerned with the larger implications of an increasingly pervasive surveillance society is not a one-off question of how performance might counter surveillance in a general sense but rather how the technologies of surveillance perform. Given the fast-paced evolution of those technologies, this is a question that must be asked again and again by artists involved in antisurveillance activism. While one can obviously debate the merits of particular strategies of agitprop activism, one thing is certain: at the time of the Surveillance Camera Players' founding the technologies behind CCTV cameras were generating alternative and far more powerful modes of performance than those espoused within the Surveillance Camera Players' own repertoire.

Some sense of those modes of performance are identified in recent discussions of the increasing prominence of biometric analysis in surveillance techniques. Although scholars like Margit Sutrop and Katrin Laas-Mikko never address the question of performance per se, they do implicitly point toward the performative dimensions of surveillance technologies in their article "From Identity Verification to Behavior Prediction: Ethical Implications of Second Generation Biometrics." In that article, Sutrop and Laas-Mikko trace recent developments in digital surveillance software that have allowed the use of biometrics to move from mere identification methods to methods that predict an individual's behavior or intentions. They note that the latter objective, which is a primary focus of second-generation biometrics, is by far the more ambitious and that it aims at providing "an automatic interpretation or decision about a person," an interpretation that is based upon a "predetermined characteristic that [ostensibly] justifies placing the person in a category of those who pose a potential threat or risk to the society."[45] Whereas first-generation biometrics focus on "characteristics readily visible to the naked eye or ear such as the iris, face, hand geometry, shape of the ear, and voice," Sutrop and Laas-Mikko note that "second generation biometric systems are focused on more intricate behavioral patterns, as indicated by gait or movement of the body or by biological traits, states, and conditions of the body (e.g., heat, smell, electrocardiogram, or DNA), with the aim of profiling people on the basis of prediction of their actions and behaviors."[46]

Equally significant in the transition from first- to second-generation biometric systems is a radical departure from the social mechanisms of panopticism. The Foucauldian notion of internalizing a panoptic sensibility is no longer necessary. While individuals are still the objects of surveillance, their awareness of surveillance is a less and less significant factor in social control. Not only are surveillance cameras now regularly equipped with "advanced software" that is "designed to perform face recognition, to capture walking behavior, scan the iris, analyze movement behavior, speech, or activities in a public space," but the cameras themselves are also decreasing in size while simultaneously increasing in the scope of their abilities to zoom and survey from distances far enough removed from those under surveillance that "one may not be aware of the fact that one's biometric features are being collected and analyzed."[47]

While there is cause for all of us to be concerned generally about the increasingly sophisticated technologies of surveillance that have accompanied second-generation biometric systems—technologies whose development has outpaced our ability to come up with responsible legal regulation of them—those same technologies ought to be of special concern to performance studies scholars and activists. For inasmuch as those technologies generate models of behavior prediction, second-generation biometric systems are performative in nature. Not only do they gather basic information and data about individuals, but that information also supplies the material and substance for the performances that second-generation systems were designed to provide. Here the multiple performances of biometric surveillance systems—their surreptitious gathering of data, their programmed assessment and then categorization of that data—lead to what, in one context, Sutrop and Laas-Mikko call the prediction of actions and behaviors but also to what, in other contexts like military defense analysis, scholars have called "performance modeling" and "performance prediction."[48] In short, contemporary systems of surveillance are actively engaged in the production of potential performances as well as in performance analysis.

While all of us presumably engage in our own acts of performance prediction when we size people up, guessing what they might do based upon the conclusions we draw about their appearance or behavior, the difference here is the sheer size, scope, and speed of the performance predictions that surveillance technologies are capable of doing on a massive scale. Those predictions are themselves based upon complex algorithmic formulas whose own performances—performances in the sense articulated by Jon Mckenzie in *Perform, or Else* (2001)—have a profound effect on the regulation of public spaces. Instances of performance modeling and prediction in the technologies and techniques of surveillance bring us into territory largely uncharted by performance studies, particularly with regard to predicted performative acts and the construction of identity. As Sutrop and Laas-Mikko note, those who assess the automated performances and performance predictions of biometric systems "are not so much interested in who you are but which kind of person you are, [and] whether you have 'hostile' intentions or malintents."[49] The interest here thus is not in the particulars of a person's identity but in what a person's voluntary and involuntary behavior suggests that he or she may intend to do.

Performative Acts and the Construction of Abnormal Behavior

The distinction between voluntary and involuntary behavior is significant. Consider, for example, the information contained in a partially declassified study on "automatic detection of abnormal behavior and threats in crowded spaces" or "ADABTS," that the EU commissioned in 2007 and that continued until mid-2013. Ostensibly, this study aims at providing "a list of identified behaviour indicators of a threat, and behaviour models for scenarios" that the EU Commission believes can assist in "the development of . . . automated detection system[s]" in public places like airports, town centers, and sports facilities.[50] While the systems of surveillance discussed in this report include systems that do "eye tracking" and "verbal protocol analysis" and assess other observable physiological characteristics like "respiratory" and "heart" rates,[51] so too do they focus on voluntary behaviors like "entering secure areas," "erratic movement," "loitering," "dropping bags," and "falling down."[52] These presumably are all generic indicators of potentially threatening behavior models.

The ADABTS report begins to move in problematic directions when it starts singling out supposed indicators like "'puffer' coats," "track suit bottoms," "designer training shoes," "baseball caps," and "ponytail hairstyles." In the aftermath of the killing of Trayvon Martin by George Zimmerman in the United States, however, other named indicators are perhaps even more problematic, especially indicators like "youths too young to drive walking through car parks with hoods up."[53] Yet even the publication of these supposed indicators is a bit of a public relations ruse. Since for "security reasons" very little of the actual ADABTS study has been declassified, one can't help but speculate that the supposed indicators of abnormal behavior and threat also include ethnic and cultural markers that the EU decided were too politically controversial to publish openly. In fact, the report comes very close to admitting as much in its "User Needs" section, which is one of the few sections published in its entirety.

A key assertion in the ADABTS report is that performance prediction software circumvents bias in assessing threats. Unlike "human police officers," automated data processing is not hampered, the report claims, by "personal interest" or "prejudices." Yet that same report not only acknowledges that "programmers may have a bias which leads to biased programming or training," but also that the identification of such biases in programming is a daunting task, particularly with regard to the more advanced systems of performance and behavior prediction software that are capable of learning and evolving. As the report concedes, it is extremely difficult for "trainers" to identify all of the "steps within the automated learning and decision-making" processes in high-end software that, for example, utilizes "Artificial Neuronal Networks" or "other forms of artificial learning techniques like . . . evolutionary computing."[54] It is amid this acknowledgment that the multiple layers of performance hovering around performance prediction software begin to come into focus.

Not only does the software predict performances and behaviors, but the performance of the software itself depends upon adequate training that is also performance-based and that ostensibly is conceptualized as a strategy for avoiding bias and bigotry in identifying indicators of potentially threatening behavior. Without any sense of irony whatsoever, for example, the ADABTS report suggests that those who "train" systems that use artificial leaning or evolutionary software should employ "(staged) training situations with deliberate variety in characteristics."[55] More than a mere suggestion, these actual staged training sessions are nothing short of the surveillance industry's own version of the Surveillance Camera Players—the difference being, of course, that these player-trainers support rather than challenge advanced forms of CCTV. Like the Surveillance Camera Players, the trainers' performances—that is, their "(staged) training sessions"—officially seek an end to racial profiling. But however well intended the report's reference to such player-trainers might be, their existence begs the more fundamental question of why evolutionary software would need to be "trained" so as not to indulge in racial profiling in the first place. At one level, then, the underlying assumption in these training sessions is fundamentally problematic and reinforces Western, white norms in appearance and behavior. The working assumption is that the only way that surveillance industry can compensate for the basic "abnormalcy" and threat of ethnic and racial diversity is by orchestrating and performing fictional scenarios in front of advanced behavior recognition cameras. Otherwise, the automated data processing systems would identify race and ethnicity as indicators of potential threats or as indicators of abnormal or antisocial behavior.

What we do know is that such bigotry and prejudice are never innate. They are passed on, taught, and learned from one generation to the next. They are programmed into our institutions and, in this instance, into our technologies. They thrive in the shadows, behind the mask, and beneath the cloak of anonymity. They also thrive in the inaccessible programming script of surveillance technologies. More important still, they thrive so as to sustain particular kinds of economic interest and privilege. Therein lies the crux of the "learning problem" with advanced surveillance software programming. The problem lies not only in the obscurity of its operation but in the obscurity of its production and marketing—in the obscurity of the economic interests, the privileges, and the class system that the supposed indicators of threat, abnormality, and antisocial behavior ultimate protect.

If that problem leads back to the entrepreneurial foundations of the surveillance technology industry more generally, so too does it lead back to the surveillance technology industry's stake in the ADABTS study in particular. While the presence of bias in the supposed indicators of threatening or abnormal behavior touches upon the kind of problematic social sorting issues that surveillance studies scholar David Lyon has discussed at length,[56] the more politically pertinent question about the ADABTS study has to do with the fact that it was commissioned rather than actually conducted by the European Union. In this respect, private enterprise has not only had a vested interest in the ADABTS study's outcomes. It has also had a seat at the table where those outcomes are decided.

Total Performance and the Surveillance Industrial Complex: A Conclusion in *Amnesia*

Those at the table included prominent representatives of the European surveillance, security, and defense industries. Companies like DETEC AS,[57] TNO,[58] FOI,[59] and BAE SYSTEMS[60] all contributed to the ADABTS report, supporting and conducting the research that led to its findings. Consequently, it ought not be shocking that the report placed little emphasis on the need to keep the responsibility for the assessment of threat, abnormal behavior, or the determination of their indicators squarely within transparent government programs that are subject to oversight and scrutiny. To do so would have been contrary to objectives of these companies. Nowhere are those objectives more clearly evident than in BAE SYSTEM's own stated mission: "To deliver sustainable growth in shareholder value through our commitment to Total Performance."[61]

What "total performance" means is not particularly difficult to comprehend. It means cultivating an enduring need for CCTV surveillance technologies as part of a strategy for maximizing profits. It means that companies that have been commissioned to assess abnormal behavior and potential threats consider themselves to be answerable first and foremost to their shareholders rather than to citizens or to government organizations. They work not toward the elimination of threat but rather toward "sustainable growth"—in other words, toward the cultivation of a permanent sense of danger by that which they have a vested interest in constructing and then identifying not only as abnormal, antisocial, and threatening but also as always elusive and thus necessitating further investments in new surveillance research and technology. Ultimately, "total performance" means the opposite of eliminating threat. It means sustaining the need for and hence increasing the profits reaped by what activists like Jay Stanley of the ACLU have called the "surveillance industrial complex."[62]

In the final analysis "Total Performance" points toward a process of surveilled acts—performative acts by normal citizens—out of the digital record of which companies like DETEC AS, TNO, FOI, and BAE SYSTEMS in Europe or ACXIOM, CHOICEPOINT, and LEXISNEXIS, in the United States surreptitiously construct identities that as Stanley notes, "are largely invisible to the average person, but make up an enormous, multi-billion-dollar industry."[63] It is in this context that we might constructively return to the Surveillance Camera Players' production of *Amnesia*, which I mentioned at the beginning of this chapter. That piece, which was borrowed from the Australian performance artist Denis Beaubois, involved a solitary performer who would suddenly appear before a CCTV camera with a board stating, "I HAVE AMNESIA." That initial board was then followed by boards stating, "YOU ARE WATCHING ME." "YOU HAVE BEEN WATCHING ME ALL DAY." "EVERYWHERE I GO." "MAYBE YOU CAN HELP." "WHO AM I?" WHAT'S MY NAME?"[64] At a qualitative level, this moment of amnesia may not be that far removed from the power and effect that new surveillance technologies may have. In the performative acts that they facilitate, acts that are cast unknowingly into sorted categories over which we have little if any control, those technologies literally establish who we are and assign us identities of which we have no recollection because those identities are stored in memory—on memory chips—to which we have not been granted access.

Data Doubles, Pepper's Ghost, and Zombie Categories: The Specters of Performance in the Bit Parts of Surveillance

> Mr. Philipp made his honors class read this Russian director-actor guy named Stanislavski, and if you were in a play, he made sure everyone knew *there are no small parts, only small actors,* like that quote was supposed to make you feel better about being stuck with a crappy role.
>
> —Charlie Quimby

Of Cosmas and Damian, or the Miracle of Body Parts as Metaphor

Sometime in the third century after the death of Jesus, two brothers by the name of Cosmas and Damian converted to Christianity. The two lived in the Roman province of Cilicia, located on the Mediterranean coast of what today is southern Turkey. According to legend, the two were martyred by beheading after torture could not compel them to renounce their faith. What is remarkable about the legend of these two brothers is that the depth of their faith was apparently matched only by their skill as physicians. Indeed, that skill was so great that their alleged medical accomplishments were deemed miracles. They were subsequently canonized and ultimately became the patron saints of medicine. Over forty-eight miracles were credited to the brothers, including the development of remedies against the plague, scabs, scurvy, and even kidney stones.[1] But their most famous miracle involved the reported replacement of a diseased leg of a Roman nobleman with the leg of a recently deceased Ethiopian or, by some accounts, with the leg of one of the nobleman's slaves.[2] This act is often depicted in paintings, and considering the fact that the first documented successful leg transplantation occurred in July 2011, Cosmas and Damian's feat was a miracle by almost any measure, especially since it was apparently accomplished without the vast array of immuno-suppressant drugs that are currently necessary to keep the body from rejecting new limbs.

The legend of Cosmas and Damian's feat may seem an odd point of departure for a study of biometrics, dataveillance, and the loss of democratic oversight amid the increasingly sophisticated global reach of surveillance technologies. But if we reexamine that legend not from the presumably divine perspective of Cosmas and Damian's miraculous accomplishment but rather from the earthly perspective of the unidentified Ethiopian without whose leg their miracle would not have been possible, then we quickly discover a structural analogy that bridges the long temporal divide separating third-century legends from the twenty-first-century realities about biometrics and dataveillance. There is, of course, the plausible argument that the Ethiopian in the miracle has always been more of an artistic contrivance than a divine inspiration, since the dark skin of the Ethiopian serves as indisputable visual testimony that the miraculous leg transplant actually occurred. Here a little color is a godsend for painters depicting this famous miracle. More interesting, however, is the question of the conditions under which Cosmas and Damian procured the leg in the first place. Somehow one cannot help but suspect that the Ethiopian (or as the case may be, the nobleman's slave) was not consulted, that his consent was never sought, and that his leg followed a surreptitious if not altogether dubious path into the otherwise saintly hands of the twin physicians. Somehow one cannot help but wonder whether there wasn't a bit of a Mengele in the miracle—that heinous "Angel of Death"—hidden in the shadows cast by the bright aura of saintliness. For all of this begs the larger question of why the Christian God would sanction Cosmas and Damian's use of their divine skills to replace the leg of the white nobleman rather than to save the life of the Ethiopian. Apparently, God does work in mysterious ways.

By way of analogy, the stolen or confiscated body parts in the mystified legends of Cosmos and Damian arguably have a lot in common with the bits of personal data that are gathered about us largely without our knowledge or consent as we make our way through public and private spaces that are increasingly regulated and defined by advanced digital surveillance technologies. I am not the first to draw this analogy. It is embedded, for example, in the structural logic of Manjula Padmanabhan's 1996 futuristic play *Harvest*, which not unlike Cosmas and Damian's miracle concerns a privileged ruling elite's dubious appropriation of body parts from a disenfranchised cultural Other. But whereas Cosmas and Damian ostensibly acted as the agents of a watchful but distant God who surveils from divine heights and with divine purpose, Padmanabhan's *Harvest* addresses the godlike advantages that sustain first-world/third-world disparities in the global marketplace. The play is populated by guards from a mysterious private company called InterPlanta Services, who act as agents for a distant privileged coterie of individuals maintaining their economic and political power through an astonishingly pervasive and watchful privatized surveillance scheme. Here personal data and body parts blur into transportable commodities. Here too privacy falls prey to the coercive logic of the marketplace, where jobs are scarce and personally debilitating compromises are an unavoidable necessity for economic survival.

Harvest, the Rules of Right, and Dataveillance

First produced in 1997 at the Teatro Texnis in Athens, Greece, Padmanabhan's play is set in a futuristic society in the year 2010, which, following the late twentieth-century logic of the play's sci-fi speculations, is a time when technological advances and the processes of globalization not only have significantly extended the reach of first-world businesses into the daily lives of the third world's disenfranchised poor but have also allowed those businesses to traffic in suspect commodities like body parts and to do so unhampered by local, national, or international police authorities. While much of Padmanabhan's play can be read as a reflection on the perverse economies that regulate international trade in human organs, *Harvest* is perhaps much more significant as a visionary work that at a very early date grasped with subtle but amazing clarity the digital revolution that during the short decade and a half after the play's premier would radically transform how public and private spheres perform. Without yet having the words to identify what it sketches as a pending reality, *Harvest* pointed toward what surveillance studies scholars now routinely refer to as biometrics and dataveillance, both of which have profoundly significant performative dimensions that I want to discuss at some length. But before moving into that discussion, I want to clarify some of the crucial terms and conditions that make dataveillance possible and thereby place the significance of *Harvest*'s creative speculation about emergent surveillance technologies in critical historical relief.

The play takes place in a small, single-room apartment in the tenement slums of Bombay that is shared by a young woman by the name of Jaya, her husband Om Prakash, her mother-in-law Ma, and her brother-in-law Jeetu. Though long unemployed, Jaya's husband Om miraculously finds himself selected from some six thousand other candidates for the very special kind of employment that InterPlanta Services offers: Om will be financially compensated for agreeing to be a living donor, held in reserve for any body parts that a receiver/client from North America might need in the future. The excessive intrusiveness and the loss of privacy that this arrangement ultimately entails becomes quickly evident shortly after Om returns from the successful interview, and the Prakash home is invaded by representatives from InterPlanta Services, who sanitize the apartment, provide the Prakash family with a six-month supply of a specialized diet that the entire family must accept, set up what is described as a contact module, and finally make sure that Om understands what in another context Graham Sewell and James Barker have called "the rules of right."

At one level "the rules of right" represent some of the most accepted forms of surveillance in society, and they provide what is the bedrock of much of the data that private corporations and government organizations collect about us—data that in many respects renders visual forms of surveillance obsolete. The rules of right are not so much the codes of conduct that one agrees to abide by as an employee, like arriving and departing at certain times, adhering to certain dress codes, or agreeing not to fraternize with subordinates. Rather they refer to a willing submission to systems of surveillance and oversight: like pilots agreeing to mandatory drug testing; customer service operators agreeing to a monitoring of their phone conversations with customers;[3] curriers agreeing to GPS tracking devices attached to their assigned vehicles; bank employees agreeing to polygraph tests; computer operators agreeing to keystroke monitoring software on their computers; government employees and contractors agreeing to an entire regimen of biometric profiling like finger printing, iris scans, and DNA sampling as well as to a close monitoring of their social contacts, their travel, and their bank accounts. Of course, this doesn't even touch upon the seemingly voluntary relinquishment of privacy that individuals offer on a daily basis in order to have access to services that they gradually come to take for granted. Here the rules of right necessitate exchanges like individuals obtaining access to free Gmail accounts by tacitly allowing Google to monitor the content of their emails for a whole host of dubious clients; mobile phone users implicitly granting companies like AT&T, Verizon, O2, and T-Mobile the right to use cell phones as tracking devices for the police, the FBI, the NSA, the British Home Office, and other government agencies; social media users gaining access to Facebook by allowing the company to sell personal data to advertisers; shoppers obtaining small discounts on their purchases by allowing the retail industry to profile their buying habits as a marketable commodity; and, would-be travelers gaining access to airports by willingly submitting not only to full-body scans and searches but also to profoundly unscientific and biased forms of racial, ethnic, national, and economic profiling and sorting.

Simply put, refusing to accept the rules of right is likely to forestall if not preclude employment and access to a whole host of areas that not only are essential to daily life but also define us as functioning members of society. Ultimately, the rules of right have a profound effect on the kinds of performative acts that shape individual identity, and those acts are not only deeply entrenched in processes of social sorting and stratification; they sustain those processes as well. Sewell and Barker argue that the "rules of right" represent ways of "shaping our behavior to conform with . . . power relationships"[4] since they are voluntary in name only. But what does it mean to say that they are voluntary in name only? At one level, it is to suggest a not always subtle form of coercion, forcing decisions that are masked beneath an equivocal political rhetoric that gives them the appearance of having been the product of individual choices when in fact they are the dictated terms for continued access to the means of cultural, political, and economic survival.

In the larger equation of conformity with existing power relationships, the trade-offs necessary to gain access to things like social media might rightly seem to be a rather trivial exchange when compared to the more profound compromises that the rules of right often demand. In their most grotesque form, such compromises are often a matter of life and death, and they frequently reaffirm the power relationships that sustain global, political, and economic divisions. Indeed, they often literally make the science fiction fantasies of Padmanabhan's *Harvest* appear to be a quant form of realism. The perverse logic of these compromises—compromises that are voluntary in name only—are certainly on display in the dramatic portrayal of the financial arrangements that Padmanabhan's character Om makes when he decides to serve as a living donor so as to provide for his family. It is an act of desperation performed under the guise of a voluntary choice, and that desperation eclipses, indeed precludes, the pursuit of anything that might resemble genuine volition.

At first blush, this arrangement literally sounds like the stuff of futuristic dystopian fiction, and, in fact, scholars like Shital Pravinchandra have argued that Padmanabhan's "play confronts us with a futuristic Bombay of the year 2010, a time," so argues Pravinchandra, "when legal, moral and bioethical debates about organ transplants have been overcome."[5] But in the late 1990s such debates were already meaningless bits of inconsequential political theater. The debates had long since been overcome by the powerful sway of the global black market. Despite the passage of legislation in India's parliament in 1995 "banning payment for organ donation," the practice continued (and to this day persists unabated) in the communities of India's economically disenfranchised.[6] Evidence from the 1990s of this practice is not difficult to find. Discussing practices at "at the clinic of Dr. K. C. Reddy," whom she describes as "India's most outspoken advocate of the individual's 'right to sell' a kidney," Berkeley anthropologist and international human rights activist Nancy Scheper-Hughes recounts, for example, the desperate plight of countless women in the 1990s from the most distressed communities of India's vast economic landscape, women whose desperation precipitated acts and behavior that have not only brought them into conformity with, but have also made them participants in, power relationships that survive through their own seemingly voluntary submission to debilitating forms of exploitation. In a manner wholly consistent with the images emerging in Padmanabhan's play, the women cited by Scheper-Hughes belong to the many who have "voluntarily" traded one of their kidneys for financial remuneration. Writing specifically about women from the Chennai slums in South India, Scheper-Hughes notes:

> The kidney sale was usually preceded by a financial crisis: the family had run out of credit and the bill collectors were at the door. The women said that the money from selling their kidneys had offered temporary relief but that it was soon swallowed up by the usurious interest charged by local moneylenders. The families were all in debt again. They also stated emphatically that they would do it again. If only they had three kidneys, with two to spare, then things might be better.[7]

Given the direness of the women's situation, Scheper-Hughes rightly suggests that it is disingenuous to speak of volition or of an individual's "right to sell" when decisions are the product of a desperation that existing power relationships do more to sustain than to alleviate. By every reasonable measure, the women mentioned by Scheper-Hughes were compelled by circumstances beyond their control and by the harsh rules of an economic order that cynically defines the notion of an individual's "rights" as a self-destructive submission to exploitation within the dark economies of the global medical black market.

Of Love, Affection, and Devoted Data

Further examination of the unsavory market for paid transplants in India gives some indication of just how far this kind of self-destructive submission actually extends. Shortly after Scheper-Hughes published her essay "The Ends of the Body: Commodity Fetishism and the Global Traffic in Organs," Vivekanand Jha pointed out that the 1995 Indian Organ Transplant Act included an exemption clause, which allowed for the harvesting of organs from live donors on altruistic grounds. While this exemption supposedly did not allow for financial remuneration, Jha notes that "in some states, Committees approved thousands of paid donor transplants after getting affidavits from donors stating that the donations were being made on the grounds of 'love and affection' for the recipients." On the face of it, it would be hard to disagree with Jha's assertion that no one can seriously believe "that so many poor, illiterate, out of work strangers developed enough affection for rich recipients from far off places whom they have either never met or just seen a couple of times to be able to donate an organ for them."[8]

Following this line of argument, the "love and affection" affidavits might be dismissed as little more than a perverse charade designed to mask a late twentieth-century slavery narrative that is reminiscent of all of the darker undercurrents of the saintly legend of Cosmos and Damian and the myth of the Ethiopian slave who supposedly out of a sense of devotion willingly offered his leg as a transplant for the diseased leg of his master. But at a subtler level so too do those affidavits suggest a path between the rules of right that cultivate live donors and those that ultimately facilitate the harvest of multiple forms of personal data. In both instances, the rules of right function solely because those who submit to them are convinced of the necessity of obtaining what they receive in exchange for what they sacrifice. Although one must be careful not to trivialize the grotesqueness of the global trade in body parts and the life-threatening exploitation of donors that this trade involves, it is worth considering whether those affidavits serve a larger exemplary function than merely providing dubious medical merchants with a convenient legal circumvention. At the very least, the affidavits cited by Jha give evidence of a desperation so extreme that those caught in its snares willingly provide wealthy recipients not only with body parts but, in those affidavits, with a kind of prepackaged moral dispensation for being at the receiving end of an otherwise ethically indefensible economy of medical cannibalism.

There is illusion in that dispensation, of course, but there is little illusion in the larger function that the affidavits also serve by providing recipients with legal coverage that technically exempts them from liability or culpability in a suspect medical practice. As is consistently the case with the operative structures of the rules of right, those written expressions of "love and affection" provide a paper trail that can be read in court as expressions of volition. They are requirements of a legislative act—the Indian Organ Transplant Act—that ostensibly protects the disenfranchised while in effect actually providing the wealthy greater legal protection than they previously had.

An identifiable resonance echoes between this kind of legal trickery and the exercise of the "exceptionalism of power" that I mentioned in chapter 3 and in my discussion of how the operative structures of the ban-opticon circumvented actual accountability for the human rights abuses at Abu Ghraib and Guantánamo. But here a demand that donors sign a document legally and morally absolving organ recipients from culpability is only an extreme example of the terms of service—the rules of right—that each of us signs off on countless times any given year. In exchange for the kind of access, service, or remuneration that enables us to be functioning members of society, we all routinely and enthusiastically provide wealthy corporations not only with compromising pieces of personal data but also with the legal basis to do with it what they will.

How do we account for such enthusiastic relinquishment of one's private information into the hands of strangers? How do we explain the irrational trust that we place in the corporations that harvest data about us, which they then in turn repackage as a marketable commodity? The answers are here multiple, but perhaps the irrational expressions of "love and affection" by live organ donors in India point toward a deep internalization of existing structures of authority and power rather toward a mere legal circumvention, and in this respect perhaps those expressions are not so unique and thus point toward the structures of authority that are also at play in our casual acceptance of the increasingly invasive tactics employed by the large corporations that collect information about us. I want to look at the notion of internalization more closely, but it is important also to note that it frequently intersects with a general lack of understanding of how data is actually used in an era of data processing that even Padmanabhan in her futuristic drama *Harvest* did not foresee.

First, consider the question of internalization: What is perhaps most disturbing in Jha's account of the organ trade in India is that it overlooks the deep sense of identification with power that is evident in the very statements of "love and affection" that, for understandable reasons, he urges us to dismiss as implausible nonsense. I want to suggest by contrast that there is much to be gained by giving those statements credibility and by considering the ways that expressions of "love and affection" for powerful strangers—expressions ultimately manifested in self-destructive acts—resonate across history and culture to recall the phenomena of "altruistic surrender" and "identifying with the aggressor" (or with one's oppressor) that was first explored by Anna Freud and then by Bruno Bettelheim.[9] Both involve a process whereby those who find themselves trapped within the discursive logic of a social, political, cultural, or emotional order begin to internalize and affirm the values of that order as a mechanism for survival—even in situations where the affirmation of those values ultimately proves to be self-destructive. Perhaps the most extreme examples of this tendency are to be found in Bettelheim's disturbing firsthand accounts of prisoners at Buchenwald who were so psychologically traumatized by their horrific experiences there that they gradually began to imitate their Nazi oppressors. Less extreme but more frequent examples are to be found in the numerous ways that the poor identify with the ruling wealthy elite in the hope that this will somehow ameliorate their own exploitation and dire circumstances—the kind of identification that arguably got Donald Trump elected president.

Just how far removed we in the West are from such tendencies in our readiness to acquiesce to "the rules of right" that facilitate the collection of massive amounts of personal data is difficult to pinpoint, but given the astonishing level of public indifference to the revelations that Edward Snowden made of the NSA's blanket surveillance of citizens, for example, it would seem pretty evident that the public has a deep attachment to if not an outright identification with the very technologies that make it possible for the NSA and its numerous private contractors to accumulate the kind of personal data on virtually every single person that renders them vulnerable to manipulation and control should powerful interests see them as a threat.

A prescient critical engagement with this latter kind of identification with the aggressor took center stage in Padmanabhan's play *Harvest*. It is most evident when the character Jaya challenges her husband Om's naive trust in the company InterPlanta Services and questions in particular his devotion to "Ginni," the Western recipient-client whose image is projected in a proto-Skype-like fashion into their apartment through the contact module that InterPlanta Services has installed. Mistakenly thinking that Ginni can only see them when they see her, Jaya scoffs dismissively at InterPlanta Services and criticizes her husband, saying, "What faith you have in them! They don't care about any of us, not as people, not as human beings." In response, her husband Om counters "What're you saying? You don't talk enough to Ginni. If you did, you wouldn't feel this way. . . . She really cares for us."[10] There is little question in the play that Jaya is right, and that Om's devotion to Ginni is a cultivated product: the result of a desperation extreme enough that he is largely blind to the destructive consequences that his actions have not only for him but for his family as well. But what neither of them understands is that Om's identification with Ginni implicates all those around him. Om's body organs prove to be a secondary matter of concern and a mere pretext to set in motion a surveillance program designed to ensnare his bother Jeetu and through Jeetu ultimately to go after Om's wife Jaya, who, as InterPlanta Services already knows, is secretly having an affair with Jeetu. The implication here is that InterPlanta Services has secretly gathered information on Om's wife and his brother through a larger unspecified surveillance net well in advance of selecting him out of the other six thousand who had applied for the job that he ultimately received. This is, as Jaya later learns, why Om was offered the job in the first place.

Almost all of us have experienced some form of "the rules of right" in our working lives, and much to her credit Padmanabhan recognized earlier than most how acquiescence to those rules tends to spill into areas well beyond what those who sign off on them actually expected or could perceive would be the case. Beyond the sense in which signing off on "the rules of right" involves a tacit identification with authority, this spilling over of the collection of personal data into other areas that remain unknown to the individual whose data is passed around as marketable commodities is perhaps more alarming than the kind of identification with the aggressor that is exemplified in the character Om's irrational devotion to "Ginni." But while Padmanabhan had some premonition of this spillage, she could only point vaguely at what it has now become and at the dynamic way that it actually functions. The implication in the play is that the ultimate interest in Jaya is the product of an all-encompassing regime of surveillance that functions through clandestine forms of observation. In point of fact, the gathering and processing of personal data follows a dynamic that is substantially more alarming because it is more difficult to pinpoint and regulate.

In the case of Om in Padmanabhan's play, acquiescence to "the rules of right" unknowingly commits, without their expressed consent, the members of his immediate family to a system of surveillance that penetrates virtually all aspects of their daily lives. Much of this surveillance occurs through the contact module that InterPlanta Services has installed in the Prakash home. But as we learn later in the play, the contact module does much more than merely serve as a high-tech video communication device. It hovers about the Prakash flat like a small drone and constantly collects data about the entire family: data like biometric information on the "skin and blood"[11] of Om's brother Jeetu and of Jaya herself, whom the corporate figures of InterPlanta Services then try to strong-arm into being a surrogate mother. What is noteworthy in this scenario is also what is perhaps most easily overlooked. Despite her otherwise perceptive vision into the future, here Padmanabhan actually reinforces popular conceptions of how corporations collect and use personal data. The image of the module that she presents us with assumes a kind of one-to-one relation between data collected and knowledge gained.

Beware of Big ~~Brother~~ Data!

The problem is that the processing of data by private companies no longer functions this way and hasn't for some time. What Padmanabhan failed to recognize is how rapid technological advancements have radically increased the ability of corporations to collect and process massive amounts of seemingly unrelated personal data, and these powerful advances in data collection and processing have opened the door to actual possibilities that in their potential literally surpass the dark dystopian vision that Padmanabhan offers in *Harvest*. Some sense of the world that Padmanabhan was unable to foresee in the late 1990s can be garnered from Viktor Mayer-Schönberger and Kenneth Cukier's recent book *Big Data: A Revolution That Will Transform How We Live, Work and Think*. In their pathbreaking study, Mayer-Schönberger and Cukier suggest that such one-to-one notions of data and knowledge are grounded in the logic of causality when in fact the revolution occurring in the era of what is commonly called "big data" involves a shift to what they call "the logic of correlation." The key moment in the big-data revolution, they argue, came when corporations realized that collected data had seemingly inexhaustible potential if it were "no longer regarded as static or stale" and if those collecting data about us abandoned the notion that data's "usefulness was finished once the purpose for which it was [originally] collected was achieved."[12] In an era where the memory capacity of computers is increasing with each passing day and is matched only by the growing speed and sophistication of computers' ability to engage in highly advanced forms of data processing, Mayer-Schönberger and Cukier argue that data can be infinitely recycled and reprocessed in ways that reveal all sorts of unexpected and meaningful statistical correlations—like Google boasting in 2009 that it could accurately predict the spread of the winter flu virus in the United States based solely on their records of "what people were searching for on the Internet."[13]

Following the logic of this kind of statistical analysis—that is, the logic of correlation—big-data brokers "in the United States such as Acxion, Ecperian, and Equifax charge handsomely for comprehensive dossiers of personal information on hundreds of millions of consumers."[14] These dossiers are not static lists of facts. They are the product of sophisticated statistical analyses by highly advanced computers.

> The industry of collecting, aggregating, and brokering personal data is known as "database marketing." The second-largest company in this field, Acxiom, has 23,000 computer servers that process more than 50 trillion data transactions per year, according to *The New York Times*. It claims to have records on hundreds of millions of Americans, including 1.1 billion browser cookies (small pieces of data sent from a website, used to track the user's activity), 200 million mobile profiles, and an average of 1,500 pieces of data per consumer. These data include information gleaned from publicly available records like home valuation and vehicle ownership, information about online behavior tracked through cookies, browser advertising, and the like, data from customer surveys, and "offline" buying behavior. The CEO, Scott Howe, says, "Our digital reach will soon approach nearly every Internet user in the US."[15]

Using the logic of correlation, the processing of data by big-data brokers is emerging as one of the most important industries in the twenty-first century because it can produce astonishingly accurate predictions that allow large corporations and government agencies to assess, profile, and sort individuals according to everything from their presumed potential as a customer to the probability that they pose a security risk.

These analyses probe deep into the personal lives of countless individuals. In an eerie echo of the kind of foreknowledge that Inter-Planta Services has of the entire Prakash family in Padmanabhan's play, for example, Mayer-Schönberger and Cukier note:

> The uses of correlations are being extended even further. Aviva, a large insurance firm, has studied the idea of using credit reports and consumer-marketing data as proxies for the analysis of blood and urine samples for certain applicants. The intent is to identify those who may be at higher risk of illnesses like high blood pressure, diabetes or depression.[16]

What is noteworthy in Mayer-Schönberger and Cukier's discussion of Aviva is that an insurance company rather than, say, the World Health Organization, is looking at credit reports and consumer-marking data as proxies for actual medical examination. We should have no illusions about what this means. It would be no great leap of conjecture to conclude that companies like Aviva are less concerned with solving the problems of hypertension, diabetes, or depression than they are with finding a discrete way to deny coverage to people with preexisting health conditions. This is the dark side of big data and its logic of correlation.

In many respects, that dark side is simply part of the long shadow cast by neoliberalism and by public policy initiatives that gesture toward a reduction of all aspects of social experience into something quantifiable, that is, into processable data. This is as true in the health care industry as anywhere else. At almost the same time that India passed the 1995 Indian Organ Transplant Act, for example, the United States was drafting legislation that led to the Health Insurance Portability and Accountability Act—an act that legislated "the adoption of electronic systems for information transfer between healthcare providers and insurance companies."[17] As Jill Fisher and Torin Monahan have observed, this legislation spurred the development of technologies like RFID (radio frequency identification) for "the management of patients, personnel, and inventory."[18] The dark side of this legislation, according to Fisher and Monahan, is that such technologies represent "a new form of social control over bodies in the post-industrial era" and do much to facilitate "new methods of data collection." Not only are such methods indicative of the increasing dominance of neoliberal "audit cultures" in the workplace—where work itself is valued only to the extent that it can be quantified—but they are also rife with "surveillance modalities, meaning that the data collected by them can be used for surveillance purposes" and that they mark one of the many areas of a growing "surveillance of people through automated systems and databases."[19] Ultimately, what Fisher and Monahan remind us of is that much that can be concluded from Mayer-Schönberger and Cukier's study is a lot less optimistic about the potential of big-data processing than Mayer-Schönberger and Cukier themselves are in their book.

Regardless of whether one is alarmed or excited about the possibilities that Mayer-Schönberger and Cukier describe, one thing is clear from their careful examination of big data and the logic of correlation. Phenomena like identifying with the aggressor are not sufficient on their own as explanations for why individuals seem so willing to place their private personal information into the hands of corporations that in turn repackage it as a marketable commodity. In point of fact, our personal information is now located in our most banal, simple, and seemingly insignificant acts—acts that it would never occur to us to guard from scrutiny. Studies like Mayer-Schönberger and Cukier's indicate that the most innocent Internet searches, purchases at the grocery store, and even a daily commute on a toll road all feed data into databases where it takes on a life of its own and where, following sophisticated models of correlation, the data may ultimately determine decisions about matters with no direct connection to it. A world where simple Internet searches can predict the spread of influenza is a world where that same data can ultimately also determine corporate decisions about whether one receives adequate health care coverage or a mortgage loan, whether one is targeted by law enforcement, whether one is admitted to university, or whether one is offered a job. This is increasingly the applied potential of each bit of data collected about us. It is compiled, processed, and reprocessed on an ever-increasing scale in what has now become—and we are witness to this—a self-perpetuating, never-ending cycle.

Welcome to the world of dataveillance—a world that has literally eclipsed the ocular foundations of the surveillance society and has reoriented its priorities. Think what we will of the omnipresence of

CCTV cameras—those most visible signs of the surveillance society—they are anachronisms and artifacts of a quickly expiring world. Their function is to document, to intimidate, and, in some instances, to provide real-time feedback. But in the era of big data, they play a distant secondary role to what we now call dataveillance. In brief, this is why the Surveillance Camera Players are now defunct. The power of dataveillance lies in its ability to predict and to look beyond the past or the present into a future that it is able to predict with increasing degrees of accuracy, ironically, in large part because of the increasingly influential role that it is playing in shaping that future.

Biometric Data, Dataveillance, and Function Creep

An awareness of the new realities described by Mayer-Schönberger and Cukier in their book *Big Data* also casts an aura of anachronism over works like Manjula Padmanabhan's *Harvest* or at least over images like that of the rather quaint futuristic contact module that she has hovering about the Prakash flat transmitting data on the behavioral patterns of the entire family. And yet, here too there is much that is still correct, particularly in the anxious premonition that Padmanabhan has about things to come, even if that anxiety about coming developments resides primarily in the margins of her narrative. It is hard to find those margins amid the chaos of the small drone hovering about the apartment and the deeply disturbing image of agents forcibly removing Jeetu rather than Om from the Prakash home so as to extract his eyes for a Western client. But however much that contact module might herald the advent of private corporate drones, Om's wife Jaya finally realizes late in the play that this misleading little module (like the vast array of CCTV cameras in our cities today) is actually of minor significance in relation to the apparent violations of privacy that led InterPlanta Services into her home in the first place.

It is never entirely clear how InterPlanta Services learned about Jeetu, the love affair that he and Jaya are having, or ultimately about Jaya's suitability as a surrogate mother. But in the larger scope of the issues of surveillance addressed by the play, that lack of clarity is actually quite telling because it leaves the door open to the multiple possibilities that were just around the corner and that we now face as reality. InterPlanta Services' knowledge about Jaya could be the product of big data and the logic of correlation, advanced statistical analysis, and calculated probabilities. It could also be the product of surreptitious surveillance that may or may not have been legal but that is virtually impossible to regulate because it is hidden from public scrutiny and oversight. It could be the product of personal data that was collected, stored with lax security, and accessed for purposes other than those for which it was originally collected. It could be a combination of any of these. Indeed, this is one of the most pressing issues in the health industry today. In her article "Health-Care Data Protection and Biometric Authentication Policies," Cheryl Brown notes, for example, that not only are biometric measurements—like fingerprints, iris scans, computerized voice or face recognition capabilities—increasingly used in the medical community to authenticate patient identity and to access medical records, but that same biometric data is widely available in the marketplace today. In no uncertain terms, this means any breach of hospital database security protocols will allow virtually any company with a computer and Internet access to quickly match individuals with their health records.[20]

If all of this may have sounded far-fetched in 1997, it is the reality that we now face, and the issue is not just access to medical records: the issue is the biometric data itself and the potential ways that it might be adapted as a source of secondary information. Indeed, just a decade after Padmanabhan wrote *Harvest*, scholars of bioethics like Emilio Mordini and Sonia Massari were already sounding alarms not about Google search data but about the use of biometric data beyond the original goal of using it for purposes of identity authentication. In a 2008 article on biometric data collection that was published in the journal *Bioethics*, Mordini and Massari specifically warn against a growing "function creep" in biometric data collection, that is, "the expansion of a process or system, where data collected for one specific purpose is subsequently used for another unintended or unauthorized purpose."[21] If, as Mayer-Schönberger and Cukier have noted, companies like Aviva Insurance are using consumer-marketing data as proxies for blood and urine samples, one can only imagine—indeed, one's imagination would be tested in speculation about—the ways that biometric data might be used. At the very least, Mordini and Massari warn that beyond the purposes of identification "biometric characteristics could become a covert source for prospective medical information, allowing people to be profiled according to their current and potential health status."[22] So deep are the concerns regarding the potential misuse of biometric data in the world of big-data processing that Mordini and Massari argue that on ethical grounds we must resist what they recognize as an already existing reality:

> Biometric characteristics should be always chosen with their potential for disclosing health details taken into consideration. Medical doctors should be involved in biometric capture device design, and sensors with the effective capability to detect medical conditions should not be deployed. Second, people should be always aware of the presence of biometric sensors. This is a very difficult goal, not only because some applications are by definition covert (e.g. screening and surveillance), but also because biometric sensors are more and more embedded in ambient intelligence environments.[23]

There is something of the belated in this ethical call. Mordini and Massari are clearly warning about a train that has long since left the station, and what they do not address are the potent economic interests that power its engine. Unfortunately, history would suggest that where bioethics and business ethics intersect, business ethics and profit motives take the right of way. The global black market for live organs is a graphic illustration of this very point, and this brings us back to Padmanabhan's *Harvest*.

Whatever its sources, the biometric information garnered through dataveillance collection technologies in Padmanabhan's play generally remains a means to an end rather than an end in itself. Although in the world envisioned by Padmanabhan those technologies are vital tools in a grotesque medical economy, the information they circulate has not yet become the specialized overt commodities or traded currencies of political power that they are today. But inasmuch as the use of these technologies is connected in the play with the appropriation of organs and bodily functions that are vital for survival—and inasmuch as the physical body of the character Jeetu is at one point completely absorbed into digital technologies and becomes something akin to a holographic projection—I want to suggest that Padmanabhan points toward a future where possession and control of the data collected by those technologies are every bit as vital to survival as one's organs are. Indeed, I want to suggest that Padmanabhan points toward a world, now already realized, where possession and control of that data are ultimately *more valuable* than control over one's own body. In fact, having control over a person's personal data will do more to coerce that individual into compromising his or her bodily integrity than having control over a person's body will do to coerce an individual into compromising his or her personal data.

That is because in the vast majority of instances, the most vital personal data about us—the data that determines everything from whether we can cross a border, use an airport, board a train, drive a vehicle, receive health care, purchase insurance, procure a visa, secure housing, find employment, obtain credit, cash a check, make phone calls, or even buy groceries—is data that does not belong to us. This is true even of biometric data (particularly when organizations collect it without our knowledge). It is collected, owned, marketed, and sold by private corporations—corporations like Donnelly Marketing Information Services, which in 2006 already had "digital dossiers" on 125 million people in the United States, or Wiland Services, which in the same year was collecting data on over 215 million people.[24] Subsequently purchased by the Neodata Corporation, Wiland Services not only now claims to have "over 235 million unique consumer records" and "billions of actual transactions" stored in its database, but also claims to "know the transactional behaviors of virtually every direct responsive adult in the U.S."[25]

Returning momentarily to the legend of Cosmas and Damian, I would suggest that with regard to our personal data it is thus not too much of a stretch, metaphorically at least, to find similarities with the Ethiopian who was not consulted about the body that would be defiled, or the leg that was secretly skirted away without his permission. And if, alternatively, that Ethiopian was the Roman nobleman's slave, as is the case in one variation of the legend's narrative, I want to suggest as well that our personal data, like his amputated leg, *is us but is not ours* because it is owned by another master to whom we defer and give our consent under duress. There is a difference here, of course. That which, in terms of personal data, is us but is not ours—that which we consent to give away under duress—not only takes on a life of its own, but also ultimately determines who we are, and what we are allowed to be.

Jeetu, Pepper's Ghost, and Tupac

The fate of that Ethiopian slave is not that far removed from the fate of Om's brother Jeetu in Padmanabhan's *Harvest*. He too is taken under duress, robbed not of his leg but of his eyes, which are replaced with prosthetic implants that give him visions of a virtual reality so desperately alluring that, as Helen Gilbert has noted, "he is seduced . . . [into] consent[ing] to the next stages of what will eventually be a total body transplant."[26] Following an empty promise of receiving his eyesight back, he willingly returns with the guards to the InterPlanta Services clinic, where he is presumably murdered and where the particulars of his body are reduced to data. At this point, Padmanabhan's drama begins to enter the realms of a kind of grand allegory about the emergent age of digital reality. InterPlanta Services uses Jeetu's body to create what amounts to a holographic image that at an opportune moment is projected into the Prakash flat in a ploy to lure Jaya (who by now has been abandoned by Om) into believing that Jeetu is still alive and that somehow through his projected image she might begin her life anew with him, provided that she agrees to be a surrogate mother.

The play ultimately ends in a stalemate, with officers from InterPlanta Services pounding at the door of the Prakash flat that Jaya has barricaded after realizing that the image she has seen is a powerful digital construct, "a data double" that is of Jeetu but that no longer functions directly as a reflection of him or his actions. His data double is thus both familiar and strange. It is recognizable but is the owned product of InterPlanta Services and functions under the company's full and autonomous control. In this latter respect, the image of Jeetu at the end of Padmanabhan's play bears striking similarity to the data doubles that each of us has and that comprise accumulated personal data, which is somehow of us but not ours, and which we do not own or control but which determines who we are and what we are allowed to be, do, and become.

It is in this image that Padmanabhan proves to have been perhaps most prescient in understanding the digital global political order that was on the horizon. In what could very well serve as an incisive assessment of the implications of the final scene in Padmanabhan's play, David Lyon has noted the crucial role, for example, that data doubles play in orchestrating a global network of surveillance. "A key dimension of contemporary surveillance," he argues, "is that it may be done at a distance, using proxies such as data-doubles and fragmentary data, from transaction records to fingerprints." "Surveillance," he continues, "tends to be removed, that is, from the local, the proximate, the face-to-face."[27] Such is the world that Jaya sees before her in the data double of Jeetu that is projected into the Prakash flat. Only we can't see our own doubles. Like Jeetu, we are blind to that which influences us most because, rather than the personified image that we get in the projection of Jeetu, it is a composite of data bits, binary code, and algorithms.

In terms of sheer theatricality, there is something vaguely reminiscent of "Pepper's Ghost" in Padmanabhan's final image of Jeetu. Pepper's Ghost, as we might recall, was a late nineteenth-century theatrical illusion invented by Henry Dircks and subsequently popularized by John Henry Pepper, who used the illusion in lectures that he gave in the 1860s in London at the Royal Polytechnic Institution and that were designed to debunk the claims of spiritualists who were popular at that time. As Beth Kettelman has recently noted, Pepper's Ghost used "a large pane of plate glass and carefully controlled illumination to allow audience members to view the reflections of hidden performers" who thus appeared as ghostly apparitions imposed upon or alongside other performers who were physically present on the stage.[28] The reality of this phantasmagoric image was based in the logic of causality and in a one-to-one relation between an actor concealed in a pit beneath the stage and the actor's image that with "a specially designed magic lantern" and sheet of glass was projected onto the stage.[29]

With the advent of digital cinema, more recent examples of Pepper's Ghost have dispensed altogether with the hidden live actor. In April 2012, for example, a company called Digital Domain Media Group stunned audiences at Coachella by resurrecting the late rapper Tupac Shakur, who sixteen years after his death suddenly appeared on the stage, joining fellow rapper Snoop Dogg to do a reprise of songs that they used to do together. Like Padmanabhan's Jeetu, the Coachella Tupac was a digital construct: something that was of Tupac but that literally did not belong to him. The small bits of data that made up his image were not only more important than the rapper they outlived, but literally performed as an autonomous, distinct, and powerfully marketable commodity.

In the new technologies of dataveillance, data bits like those that comprise the Coachella Tupac are no small matter. In fact, today Stanislavski's famous admonishment that "there are no small parts" could apply as much to the data bits that are the very substance of advanced dataveillance technologies as it originally did to the bit parts that play a subtle yet vital role in theatrical productions. Playful though this slippage between bit parts and data bits might initially appear to be, I want to suggest that it highlights the importance of performance as a crucial, albeit largely unacknowledged, mechanism in the functioning political structures behind data collection and the processing of personal information by the state and by private corporations. We would do well to consider the multiple ways in which the advanced digital technologies of dataveillance parse individual citizens into manipulable data bits that when reassembled can be made to perform according to predetermined political agendas and business models.

Such reassembled "data doubles" are, I would suggest, mere specters of our actual selves—something metaphorically akin to the theatrical illusion called Pepper's Ghost but somehow frighteningly more powerful because they ultimately have reversed the generative authority of the performative illusion and render the functioning logic behind Pepper's Ghost something resembling what Ulrich Beck has called a "zombie category." In a 2001 interview with Don Slater and George Ritzer, Beck suggested that "we are living in a society, in a world, where our basic sociological concepts are becoming what I call 'zombie categories.'" He explained further that "zombie categories are 'living dead' categories which govern our thinking but are not really able to capture the contemporary milieu."[30] The underlying assumption here is that real political change requires the kind of vision that is only possible by laying those categories to rest.

Laying one of today's most dominant zombie categories to rest necessitates a radical departure from the one-to-one logic that allows us to maintain the illusion that we are the authors and owners of our own data doubles. Whereas the ghost in Pepper's illusion is a projection of an unseen actor whose actions control the image on the stage, and while the Coachella Tupac is a digitized projection of past performances, our "data doubles," I would suggest, move us into very new territory. Not only do they perform beyond our control but in fact they determine and direct how we as individuals perform in society.

Our data doubles, I would suggest, are the source of our own disenfranchisement and will prove to be the single most powerful political commodity of the twenty-first century.

Data Bits, Personal Profiles, and the Advent of the Metacommodity

As a general concept, dataveillance has been in circulation since the late 1980s, when Roger Clark coined the term to account for "the systematic use of personal data systems in the investigation or monitoring of the actions or communications of one or more persons."[31] As Daniel Solove has noted: "There are hundreds of companies that are constructing gigantic databases of psychological profiles, amassing information about an individual's race, gender, income, hobbies, and purchases. Shards of data from our daily existence are assembled to investigate backgrounds, check credit, market products, and make a wide variety of decisions affecting our lives."[32] Although in this chapter I have only been able to mention a few of the more prominent data brokers, the technological and economic trends are clearly working in their favor, and what we can expect to see in the future are partnerships, mergers, and a consolidation of the various databases that such firms possess—all in the name of maintaining the competitive edge that is offered by analyses of larger and larger sets of data.

With the advent of digital technologies, the databases of companies like Acxiom have exploded both in terms of the quantity, scope, and coordination of the data that they collect and in terms of the sophistication of their ability to mesh and merge with other databases and thus to process a vast array of highly personalized data into increasingly distinct, marketable profiles, tailored to the needs of individual clients. All of this was an outgrowth of what in the business world is known as CRM, or consumer relations management, and, initially, dataveillance was developed as a tool to help companies refine their market strategies and thus to maximize their performance and profits. "The banking, credit card and telecommunications industries" were particularly keen on information that "could give them control over transactions and information," and thus their private sector interests have been the source and impetus of many of technologies that have made dataveillance a viable possibility today.[33]

Whatever its original impetus, big data has proven to be much more powerful than its original visionaries imagined. With increasingly sophisticated digital processing technologies at their disposal, corporate executives quickly realized that these data packages and the profiles they include were powerful commodities in their own right. Indeed, they have arguably become something of a *metacommodity* because at this point they are so sophisticated in the information that they can provide that companies like Wiland Services or, more infamously, Booz Allen Hamilton are no longer trading in market information. They are trading in political power and control, and with each passing day, the value and strength of that political power increases. Here we are also witness: witness to the emergence of an entirely unprecedented political order that is all the more powerful because the public has not yet recognized, and because the political order facilitated by dataveillance has not yet realized, its full potential.

At the most obvious level, the performative political power of such databases was on full display in the 2012 election when the Obama campaign "recruited some of the most brilliant young experts in analytics and behavioral science," microtargeted specific voting groups, and, based upon their access to massive amounts of data, ran over "66,000 simulations" daily in order "to model the state of the election." Alice Marwick notes, the lesson to be taken from "the Obama campaign's tactics" is one "that is often missed in our discussions of datamining and marketing—the fact that governments and politicians are major clients of marketing agencies and data brokers."[34] But in this regard, winning elections is only a small part of the story. The larger story is about the potential for manipulation and control of groups and individual citizens.

As we have seen in the earlier discussions from this chapter, the data mining and processing that companies like Neodata Corporation or Siebel Systems sell literally determine whether an individual can or cannot function in society, and so, at a level that is hard to overlook, what CRM companies are selling are the tools with which to manipulate and control individuals, and to do so in ways that have the advantage of being difficult to pinpoint and difficult to trace back to their sources. Ostensibly, biometric identification is designed, for example, to make it simpler to eliminate identity fraud through the use of automated systems of recognition and verification that "determine the identity of a person based on physiological or behavioral characteristics."[35] But Kelly Gates has argued that the political realities of biometric identification technologies are a lot less neutral than the public hype would lead us to believe. In her article "Biometrics and Post-9/11 Technostalgia," for example, Gates argues that, beneath the surface of their identity verification function, "biometrics are designed to transfer some of the responsibility for the culturally conflicted process of identification from humans to computer systems."[36] It doesn't take much skill as a reader to figure out that a phrase like "culturally conflicted processes of identification" functions as a euphemism for institutionalized forms of ethnic and cultural bigotry, which can thus be pursued under the guise of national security and masked beneath the logic of correlation. As Gates rightly notes, there is an incredible amount of political power in such transfers, particularly over the course of the last decade.

A good portion of this political power is as much the product of historical events as it is of technological advancements, and on this note we move into an area completely unforeseen by Padmanabhan in the future that she imagined in *Harvest*: namely, the events and aftermath of September 11, 2001. Provoking a strong ideological wave of nationalism and patriotism, which has consistently provided rhetorical cover for xenophobia and bigotry, those events led CRM executives to expand the conceptual parameters of their market and thus to repackage their data-wares to appeal to the political insecurities that were the immediate product of the attacks of 9/11. Venturing into the security industry that burgeoned in the years following 9/11, CRM companies and other data brokers radically shifted the focus of their products from providing consumer profiles to providing discrete hierarchies of suspicion and threat. Following the logic of correlation rather than causality, those hierarchies lent the aura of scientific legitimacy to what in reality was tantamount to an emotionally and politically explosive variation of the questionable processes of "social sorting"[37] that David Lyon argues are indicative of surveillance systems more generally. At one level, the events immediately following 9/11 could arguably be cited as the moment when the general public bought into the very techniques that are now being used against them on a massive scale by state and private organizations.

What 9/11 provided was the opportunity in response to the threat of terrorism to spin a compelling and lucrative narrative masked not in the language of oppressive mass-scale surveillance but in the language of American superiority, and that language had powerful resonance. As Heather Cameron recalls, for example,

> Immediately after the events of 9-11 the CRM industry began to lobby for opportunities to sell their services to the government as a way of fighting terror. On 26th of February 2002, CRM industry executives including Tom Siebel of Siebel Systems Inc. appeared before the US House Subcommittee on Technology and Procurement Policy of the House Committee on Government Reform. He explained to legislators how publicly available information on one of the 9-11 suspects could have been brought together through CRM systems to build a composite picture which would have caught the attention of law enforcement.[38]

Much of the force of Siebel's successful sales pitch before Congress resided in a form of what social science scholars like Kelly Gates and Pat Gill have called "technostalgia." But Siebel's was a voice that not only appealed to a desire to rewrite the past through the lens of technology's potential. It also appealed to what Kelly Gates calls "a collective psychological need to believe that the nation was not as vulnerable as it appeared, that our technological sophistication remains unscathed and in fact would have stopped the men had it been in place."[39] The rhetoric of this appeal should not be underestimated. For it laid the foundation for a profound moment of submissive identification with authority. It implicitly suggested that individual U.S. citizens had a kind of patriotic duty to submit to individual surveillance for the collective good.

Although Cameron is right to highlight how in the aftermath of 9/11 the public discourse about "data mining and warehousing techniques" shifted from discussions about how "to build composite pictures of customers" to how "to find patterns which are claimed to constitute terrorist activity,"[40] the blanket use of data-mining techniques by companies like Booz Allen Hamilton suggests a different story. In the name of national security, the public has willingly given private companies that cater to the government the potential to target and disrupt the life of anyone who for whatever reason becomes a person of interest or inconvenience to the clients of companies like

Booz Allen Hamilton, Science Applications International Corp. (SAIC), CACI, or BAE Systems.[41] Here the possibility of "function creep" would be almost too frightening to contemplate if we did not already have recent actual examples of such abuses demanding that we do so. As digital technologies continue to increase in sophistication and capabilities, so too does the global reach and the techniques of the political power they offer, and with that the possibilities of "function creep" become almost too numerous to fathom. And this brings us to the closing scenes of Padmanabhan's *Harvest*.

Conclusion

In the final scenes of Padmanabhan's *Harvest*, we are presented with an image of Jaya holding a shard of glass to her neck, threatening suicide, while the distant client of InterPlanta Services pleads with her not to be "unreasonable." Reminding her that it is "impossible to win against" him, the Western client asks Jaya not to make him "tell the guards to force the door."[42] Having nothing left to lose, Jaya plays the only option she has left. Leaning against the door, she tells Virgil (the Western client who as part of his ploy to gain access to Jaya had masqueraded earlier as Ginni): "If you force the door, you'll push the glass into my throat," explaining further, "If I lose my life, I win this game."[43] Buried beneath this final arresting image of a stalemate achieved through the threat of suicide is the play's single largest fantasy. It is a fantasy embedded in the logic of causality and indebted to the Hegelian master-slave narrative that became the crux of Marxist theory. Threaten her though he may, Virgil must concede that Jaya has what he needs. Jaya is literally in control of the means of production, or at the very least the means of reproduction that Virgil and his like are desperately in need of because their women are sterile and have lost the ability to reproduce. Jaya's death would thus potentially mean the demise of the threatened ruling elite that Virgil represents. (All this presumes, of course, that across the vast expanse of the third world suitable surrogate mothers are in short supply.)

With Jaya hulled up in her flat threatening suicide, the final scene of *Harvest* offers a romantic image of a calculated gambit and revolt, and it is premised upon Jaya finally becoming conscious of the leverage she actually has, holding a mirror shard that, like those collected data bits, carries a potential reflection of her. Inspiring though this final image may be, it is nonetheless largely a work of fiction. The narrative is a crutch propping up an unlikely scenario where, in the final analysis, someone like Jaya has a crucial value that can force a powerful economic elite into a stalemate. In reality, such figures are few and far between. One must first have access to become indispensable. One must be an Edward Snowden walking off with the keys to the kingdom, walking off with the data with which the powerful rule, walking off with the data to which they are beholden. But, as we ought to have known from the moment Snowden opened the vault, it would only be a short matter of time before the locks were changed and Snowden would become an artifact of history.

What Padmanabhan did not imagine was a world very much like our own where the value that Jaya possesses is her simple human dignity, which cannot be quantified and reduced to data and hence is ignored by the algorithms that increasingly shape who we are and who we are allowed to be. There is indifference and brutality in the ignorance of those algorithms and their concomitant logic of correlation. Whether coming from the character in Padmanabhan's play or some political pawn who has gained momentary leverage against larger state and corporate interests, Jaya's desperate cry for the body against the onslaught of its data double is hardly an optimistic one. But there is something familiar and easy to identify and ally oneself with in such moments of revulsion. Perhaps collectively such cries might actually work miracles and mark the beginning of what subsequent historians might call the radical redistribution of data and the egalitarian democratization of data control.

The question is whether art has a viable role to play in such redistributions and, if so, whether theater and performance have a role in particular to play. The data here is not encouraging. It points to trends that indicate the formation and fortification of what critics describe as a postdemocratic era. So to ask the question of whether theater and performance have a role to play in the radical redistribution of data begs the larger question of what is performance after democracy, which is the question taken up next in the book's final chapter.

Envisioning Performance Postdemocracy: Creative Activism, (a Little Bit of) Damage, and Surveillance's Inevitable Products

> Who still accepts the idea that to found true justice is certainly worth a bit of damage on the side of those who are well-off and their servants? And who, in the official theatre, can still find some drama in evidence of these critical facts?
>
> —Alain Badiou

About Theater about Surveillance: A Preface to a Conclusion

In June 2015, I had the privilege of seeing German actor Thomas Halle perform *Ich Bereue Nichts* (*I Regret Nothing*) at Frankfurter's Künstlerhaus Mousonturm. The prize-winning piece is a collaborative project between Jan-Christoph Gockel, Thomas Halle, and Konstantin Küspert. It premiered the previous October at the Staatstheater Karlsruhe. *Ich Bereue Nichts* is a performance monologue by Halle, who assumes the identity of Edward Snowden and offers the audience what critic Georg Patzer describes as an attempt to capture history's largest surveillance scandal through "an emotional portrait" of the now famous ex-NSA contractor.[1] To this end, Halle guides his audience through performative reconstructions of Snowden's family album. Against the backdrop of an ongoing first-person monologue in the voice of Snowden himself, Halle delves into the past, constructing images of Snowden, for example, as a small child in a cowboy suit atop a stuffed horse, as a boy before his first computer terminal, as a young man with his girlfriend in Hawaii as "an alienated IT monkey"[2] (accentuated by Halle donning a rubber chimpanzee mask). All of this is, of course, a preface to the image we now have of Snowden as the whistle-blowing, tech-savvy former contractor from Booz Allen Hamilton who absconded with a treasure trove of digital files on the NSA's massive surveillance program and who ultimately sought asylum in Russia. But even this well-known image of Snowden is a complicated one for anyone who gives surveillance serious consideration.

At some level, that kind of consideration is what makes *Ich Bereue Nichts* a noteworthy piece of theater. Indeed, one of the truly magisterial moments of this performance comes when audience members finally face the Snowden they all now know—the Snowden who, in a somewhat familiar refrain, begins a quasi lecture on surveillance overreach, the need for government transparency, and the public's need to protect itself from the government's insatiable and unconstitutional hunger for information and personal data. In order to underscore this latter point, ushers at the show brought literal transparency into the theater by turning up the auditorium lights. As Halle continued his lecture, they then began distributing every kind of helmet imaginable (football helmets, construction helmets, plumbed helmets from Roman soldiers, Prussian military helmets, motorcycle helmets, modern combat helmets, and so on). Following the obvious cue, audience members then began to try on the protective headgear. The ingeniousness of this amusing moment of participation and "dress-up" was that it was calculated as a moment of misdirection. As audience members put on their helmets, laughed at each other's appearance, and drifted into distraction, Halle quietly suited up and put on his own makeshift helmet, an upside-down, metal barber's bowl. As the lights went down again, Halle stood before the audience not merely with the inverted barber's bowl on his head but also in a suit of Spanish armor complete with a jousting lance. He stood next to the large stuffed horse from earlier in the show, which he now mounted once again, creating a hybrid image of Edward Snowden as Don Quixote.

While this hybrid image was not the actual conclusion to *Ich Bereue Nichts*, it was arguably the most important visual moment in the entire piece—if only because it complicated the image that Snowden enjoys in popular culture as a kind of white-hat hacker/whistleblower. Indeed, however admirable or heroic Snowden's actions might seem, there has always been, as Halle's performance suggests, something quixotic about his single-handed attempt to fell behemoths like the NSA and GCHQ. Halle's image of Snowden as Quixote thus serves as an important reminder that, despite his courageous disclosures, Snowden did little in terms of providing normal citizens with the actual means to protect themselves from surveillance technologies. Like the helmets distributed to the theater audience, his efforts were of symbolic rather than of practical value. If a connection between Quixote and Snowden exists, this is where it is to be found. For Cervantes's delusional character can't quite tell the difference between the constructed symbol and concrete reality. He tilts windmills that he mistakenly believes to be giants because his vast knowledge of medieval romances distorts his vision and causes him to see the world through a self-aggrandizing and anachronistic lens. But part of the appeal of Quixote is that the society in which he imagines himself to be an active agent is a better, nobler, and ultimately more malleable one than the society in which he actually resides. In his delusion, Quixote envisions a society where his actions have practical rather than mere symbolic significance and where his interventions can fell giants and thus change the world. In this respect, it is worth recalling that the only thing that most certainly changed as a consequence of the Snowden disclosures were the security protocols of the NSA and GCHQ surveillance programs that he exposed. Following this line of reasoning, it behooves us to ask, where the largely symbolic consequence of the Snowden disclosures leaves Halle's theatrical Snowden/Quixote hybrid?

Amid distinctions between symbolic and practical gestures, many would likely caution not to underestimate the cultural and political value of symbolic victories. Those of us in the arts are often the most emphatic in voicing such caution. But as I begin the conclusion of this book I cannot help but wonder whether the rise of the surveillance society has not deflated the political value of symbolic victories and whether now, as a consequence, there is not something of a quixotic undertone in that clamor about symbolism. One would do well to recall that Don Quixote's delusions result from his obsession with the literature and mind-set of a bygone era. They result from his fixation on a set of values from a place that no longer exists, and in a comparable respect this is precisely the problem not only with Snowden and his idealistic sense of democracy, but, if I can bring us back to Frankfurt's Künstlerhaus Mousonturm, it is also the problem with art and theater about surveillance—theater like *Ich Bereue Nichts*. For ultimately the Snowden/Quixote hybrid says as much about theater—and specifically theater and performance about surveillance—as it does about Snowden himself. It is not just that there was something quixotic in Snowden's actions. Those actions were immensely theatrical. Halle's performance not only made this abundantly clear, but also blurred those actions with the practice of theater itself. Amid such blurring, the implicit unanswered question at the end of *Ich Bereue Nichts* is thus whether, with the advent of the surveillance society, theater—or at least theater as we know it—has become an anachronistic lens that envisions the utopian where in fact the quixotic prevails. We can debate about how quixotic Snowden was—*Ich Bereue Nichts* asks us to do the same about theater itself. It's not merely a matter of delusion. The question is whether our current theatrical forms, like Don Quixote's antiquated book knowledge, are outdated and hence ill prepared to engage surveillance's own performances in politically efficacious and valuable ways.

Performance Postdemocracy: An Introduction

Ich Bereue Nichts is certainly not the first work to leave such questions hanging, and while such questions have loomed large over almost all political theater at least since the 1960s, *Ich Bereue Nichts*'s low-tech theatricality points us in decidedly different directions, for example, than more famous works like the Builders Association's 2005 production of *Super Vision*. Back in 2005, *Super Vision* was its own tour de force, and at some level the piece is memorable because it implicitly acknowledges the need for new theatrical forms that, as the Builders Association conceived them, might address the emerging regimes of surveillance on their own technological terms. To this end, the troupe decided on a collaborative endeavor with the design and media studio dbox. Exploring issues of surveillance and dataveillance against the backdrop of dbox's impressive multimedia stage spectacle, *Super Vision* provided a technologically mesmerizing illustration of the ways that dataveillance haunts the lives of citizens around the world. Inspired in part by John McGrath's pathbreaking book *Loving Big Brother: Surveillance Culture and Performance Space*, the Builders Association's *Super Vision* interwove three separate narratives about data and identity that were presented to audiences via a combination of "virtual sets," "high-definition projections," "live, mediated and recorded performance, as well as the real-time performance of its soundscape and supporting technologies."[3] Its multiple narratives, its sophisticated use of digital technologies, and even "the title of the piece" itself combined not only in the suggestion "that we need a new way of seeing"[4] in order to grasp the multiple levels at which dataveillance impacts our lives, but also in the implicit suggestion that *Super Vision* was the kind of the theater that would enable that new way of seeing.

Not all agreed. "Despite the edgy political commitment of the Builders [Association], despite the skill and the originality" of their work on *Super Vision*, critics like Maurya Wickstrom suggested that amid the feast of special effects "the global electronic matrix" somehow escaped "deep critique," and she ultimately concluded that the piece offers no serious "political challenge to incorporate capital's deployment of technology in the restructuring of the human itself."[5] Perhaps this shortcoming was a result of a Faustian bargain forged in the Builders Association's collaboration with dbox, which took an interest in the project that was not so much political as what dbox cofounder James Gibbs described as "keeping . . . [the] studio fresh"[6]—apparently for more lucrative commercial enterprises. But perhaps too that shortcoming resulted from an inability to reenvision theater itself in light of the political urgencies posed by the emergent surveillance society and from choosing instead to dress up an all-too-familiar theater in a sparkling new technology that made it look like more than it actually was: a theater about surveillance rather than theater that challenges and disrupts it.

I don't want to undervalue the significance of didactic theater, but here at the end of this book it is time to admit that the larger political threats posed by the surveillance society demand something more than theater *about* surveillance. They demand a theater of direct engagement, confrontation, and even disruption. In this respect, what distinguishes *Ich Bereue Nichts* from *Super Vision* is that self-reflective turn in *Ich Bereue Nichts*, that is, that moment when the piece recognizes the exhaustion of its theatrical forms and writes its own obituary with the quixotic aura that it casts over its entire endeavor. But beneath that aura and beneath the silliness of distributing all of those helmets to audience members, *Ich Bereue Nichts* arguably pines for a Snowden that is more than he was and, more importantly, for a theater that does more than *Ich Bereue Nichts* implicitly acknowledges it was capable of doing itself or that theater like *Super Vision* was capable of doing before it. The longing is for a theater whose interventions are less figurative and more literal and consequential. The longing is for a theater capable of actual political disruption and change and for a theater thus conceptualized not as politically inspired art but as artistically inspired activism.

In simplest terms, the emerging signs of this mode of performance are the focus of my concluding chapter. But in advocating this kind of approach, I realize that there are alternatives that others might choose. Some sense of those alternatives emerges at the end of Elise Morrison's book *Discipline and Desire: Surveillance Technologies in Performance*, and I cannot help but be struck by the very different conclusions with which she is left at the end of her book on art and surveillance and the conclusions that I am drawing at the end of my own book. In the coda to her book, Morrison argues that she has "come to think of surveillance art as a means of writing and enacting a vital *small history of surveillance*," the kind of history that she believes is necessary for destabilizing "the stronghold on Truth and History sought by large-scale governments and sprawling-consolidations of capitalistic power."[7] There is something deeply compelling in the conclusions she draws, and though I find myself advocating a very different kind of theatrical praxis here at the end of my own book, I leave it to our shared readers to decide which path they prefer.

If the goals of this final chapter are to examine the rudimentary formation of a theatrical praxis centered around artistically inspired activism, then perhaps in Wickstrom's disappointment with *Super Vision*, one might find some bearing in the moments when she points toward the larger contexts of that praxis: contexts in which surveillance itself is but an extension and tool—albeit a powerfully transformative one—of a postdemocratic political order. Indeed, what is noteworthy in Wickstrom's critique is her emphasis not on the pervasiveness and intrusiveness of surveillance itself, which the Builders Association captures quite well, but rather on *Super Vision*'s failure to challenge "incorporate capital's deployment of technology in the restructuring" of the political subject: its failure, in other words, to challenge the use of surveillance to construct political subjects whose own actions sustain the increasingly postdemocratic formations produced not by surveillance per se but by neoliberal policies—formations that surveillance technologies cultivate and fortify as a kind of afterthought. I will address the emergence of these postdemocratic formations in the next section of this chapter, but first one final comment on Wickstrom is in order.

Implicit in every harsh or critical theater review is a yearning for a theater that did not happen, and in Wickstrom's critique of *Super Vision* there is arguably a yearning for a theater that challenges the neoliberal currents that cultivate, legitimize, and profit from the increasingly expansive reach of surveillance technologies. Perhaps I am attributing a yearning to Wickstrom's critique that is at best inchoate, but if I am, it is because I too see this as precisely the kind of theater that the larger political contexts of surveillance necessitate. Whatever its initial forms, surveillance as a contemporary sociopolitical phenomenon is essentially a corporate enterprise and less concerned today with security in a free society than with the collection of processable data that can be packaged as a commodity and marketed globally as a tool for mass manipulation, corporate security, and political control. It is part of a brilliantly conceived industry that breeds the very anxiety it is marketed to combat. The sources of this industry are diffuse, largely, as I have argued throughout this book, because they are spurred by diverse private interests and are part of an economy—a surveillance industrial complex—that is conceptualized to elude democratic oversight and regulation. For as a sociopolitical phenomenon, surveillance is, as I will be arguing in the subsequent sections of this chapter, only the most visible manifestation of a trend toward a postdemocratic society, and so it is difficult to overstate the importance of the question, what do the postdemocratic undercurrents of the surveillance society require of a viable twenty-first-century political theater? What are the dynamics of "performance postdemocracy"?

With regard to surveillance, there are a number of ways that one might describe the new political contexts of theater and performance in twenty-first-century Western societies, but to speak of "performance postdemocracy" is to place theater and performance within the larger frame of Colin Crouch's trenchant analysis of the consequences of neoliberalism in his 2004 book *Post-democracy*. Indeed, not only do I want to argue for the need to contextualize the rise of the surveillance society against the backdrop of Crouch's analysis, but in doing so, I am also arguing three specific points: first, that beneath the pretense of "business-as-usual democracy" the rules of our society have fundamentally changed; second, that the surveillance society is the fortification of that change; and third that, as a consequence, so too must theater and performance reconstitute themselves in response to that change. But what kind of theater and performance do these changes demand?

The options here are not of our own choosing. To speak of performance postdemocracy is to speak in accord with the distant but recognizable echo of Marx's famous opening to *The Eighteenth Brumaire*, where he famously argues that "men make their own history, but they do not make it just as they please; they do not make it under circumstances chosen by themselves, but under circumstances directly encountered, [and] given."[8] But whereas Marx refers to circumstances inherited from the long arc of history, the contexts and circumstances of postdemocracy and of performance postdemocracy are the product of technologies that accelerate history itself and that have catapulted the surveillance society into place in advance of any transparent political consideration of its long-term consequences. We can debate what those consequences are and ultimately may be, but, following the logic of Crouch's own arguments about the consequences of neoliberalism, I want to suggest that the emergence of the surveillance society marks not so much the circumvention of democracy as the advent of a period of postdemocracy from which there is no return, a period in which the institutional semblances of democracy remain intact despite all of its mechanisms having been thoroughly compromised by corruption and co-optation. This places very different demands on theater. We may make our own theater, but we do not make it as we please. We make it under circumstances not chosen by ourselves and on a grid whose operations and whose operators are manifestations of a new political formation that we have not elected.

Within this context, what I refer to with the phrase "performance postdemocracy" is a mode of performance that, in response to increasingly undemocratic political formations, has abandoned conventional theatrical forms in favor of direct antagonistic engagement. It is less interested in symbolic gestures than in creative acts of confrontational interference, targeted political vandalism, and antisurveillance disturbances. Its goals are not to elude surveillance, but to engage it on terms that complicate and overwhelm its processes. It does not attempt to function beneath the radar but rather to disrupt it. While there are a number of activist-artists engaged in this emerging mode of performance—as well as precursors like Critical Art Ensemble and the Electronic Disturbance Theater—I want to focus in particular in this final chapter on the creative interventions by the German artist-activist Christoph Faulhaber and the Russian dissident artist Pyotr Pavlensky, both of whom I mentioned in the introduction to this book and both of whom have targeted the postdemocratic operative assumptions of the surveillance society in their work. But before looking closer at their work, it is important to have a clearer understanding not only of the dynamics of postdemocracy but also of how surveillance functions within it.

Postdemocracy, Neoliberalism, and Surveillance

The common refrain is that the proliferation of surveillance technologies threatens democracy. The more accurate—and in many respects, the more disconcerting—reality is that surveillance merely reinforces, solidifies, and secures the political trends that scholars like Colin Crouch argue have inaugurated an era of postdemocracy. Crouch has famously argued that the most significant political consequence of the late twentieth-century rise of neoliberalism was not the barrage of policies favoring "the interests of the wealthy—those who [benefit] from the unrestricted operation of the capitalist economy rather than those . . . [needing] protection from it"[9]—but the outsourcing of government services to private contractors. Troubling though neoliberalism's massive deregulation of business and corporate enterprises might be (especially since this has undercut the government's ability to fulfill its oversight responsibilities), the more decisive factor contributing to the emergence of postdemocracy was the simultaneous movement in government "to divest itself of all direct responsibilities for the conduct of public services"[10] and thus to subordinate those services to the business models developed by private contractors. Ostensibly, this is supposed to bring business efficiency to government bureaucracies, but in practical terms divestment and outsourcing have resulted in Western governments becoming increasingly beholden to private companies, upon which they now rely not only to provide government services but ironically also to possess the competency and means to provide oversight of those services. As Crouch argues, the "more state functions are sub-contracted to the private sector," the more "the state begins to lose [the] competence to do things which once it managed very well" and the more it is thus "forced to sub-contract further and buy consultancy services to tell it how to do its own job"[11] and to do so following terms that favor further deregulation and divestment and that consequently reduce actual democratic governance and oversight to little more than a farce. More importantly, this shift in competency, according to Crouch, has permanently altered the foundational structure of what were once the basic institutions of democratic society.

Perhaps none of this divestment would matter much—or lead to Crouch's claim that we have moved into an era of postdemocracy—were it not for factors in that divestment that bring us back to the issues of surveillance that have surfaced time and again in the various chapters of this book. Two stand out in particular. At the most obvious level, there is the basic conflict of interest that always shadows private contractors who have as much—if not more—at stake in keeping their companies solvent and their shareholders happy as they do in providing effective government services. There is perhaps no more concrete example of this conflict than the grotesque logic that prevails within the private prison industry, whose lobbyists routinely oppose criminal justice reform that might jeopardize the ability of companies like the Corrections Corporation of America to maximize its profits by keeping its facilities full of prisoners. As I argued earlier in the chapter on maximum surveillance environments, this is surveillance for profit—pure and simple—and it is tolerated not only because of the neoliberal ideologies that rationalize its obscene and consistently racist exploitation of those who spend their lives behind bars but also because the alleged "security" that such corporations provide caters not to everyone, as one might rightfully expect of a democratic society, but rather to privileged segments of society. This, according to Crouch, is the more fundamental and ultimately postdemocratic problem of outsourcing, and, I would add, of surveillance as well.

Whereas democratic governance is inclusive, business enterprises are selective. In the private sector, businesses streamline their enterprises all the time so as to maximize their efficiency and profits by strategically targeting only those potential clients who are capable of delivering a meaningful return and who will thus contribute to a company's continued growth and strength. Other consumers are disregarded. Crouch argues that when government agencies are restructured according to business models—which is one of the key objectives when private contractors and consultants are recruited—this same logic overtakes agencies that historically have attended to the needs of all rather than to those of a targeted few. Discriminatory practices result, and competent government services are scaled back, attending primarily to those "who matter," that is, to the privileged and influential.[12] But how does this relate to surveillance? If together the outsourcing of government services to private contractors and the consequent selective targeting of those services to privileged segments of society produce a dangerous and undemocratic mix, that mix becomes toxic when applied to matters of state security, secrecy, and surveillance.

In the security sector, the extensive use of private contractors and consultants by the government introduces divestment into what are arguably the most sensitive areas of democratic society. The result is a climate where security and even state secrecy cater primarily to a targeted elite—large corporate interests like those of the oil industry, for example—and extend to the general population only insofar as they fall within the longer shadow of those primary concerns. Since such government functions also require specific security clearances for active personnel, when they are outsourced, legal structures further remove security contractors and consultants from any meaningful democratic oversight. But the issue here is not merely that this equation is structured in such a way that it keeps the public in the dark. The sheer magnitude of the divestment—the massive outsourcing of government services to private military and security contractors—keeps the government itself in the dark and solidifies the shift toward postdemocracy. In their book *Top Secret America: The Rise of the New American Security State*, Dana Priest and William Arkin, for example, document the myriad ways that in the aftermath of the events of September 11, 2001, the private security industry that services the government mushroomed into a complex so vast in size, so uncoordinated in structure, and so diversified in areas of specialization that no elected officials or centralized government agencies possess the wherewithal, let alone the specialized knowledge, to exercise their oversight responsibilities. Neither the right hand nor the left hand knows what the other is doing—or whose interests the other ultimately serves.

Rather than rehearsing all of Priest and Arkin's arguments, perhaps it suffices to remember that Edward Snowden himself was a private contractor and was working for Booz Allen Hamilton when he decided to become a whistleblower. He was typical of an entire class of post-9/11 contractors: he was and in fact still is what Vinca Kruk and Daniel van der Velden, founders of the activist design studio Metahaven, describe as "a sort of freelancer, a temporary subcontractor who had no permanent affiliation with the agency; a well-paid information mercenary under neoliberal flex work,"[13] who was tasked with maintaining systems that spied on the United States' own citizens. If nothing else, Snowden serves as a reminder of the extent to which, in the aftermath of the events of September 11, the new security state documented by Priest and Arkin turned its surveillance capabilities inward against its citizens as much as it did outward against its potential enemies on the international stage. Indeed, Crouch argues, in what is arguably the third and most important factor in his discussion of postdemocracy, that the events of September 11 gave U.S. and European governments "new justifications for state secrecy and [for] the refusal of rights to scrutinize state activities"—justifications that circumvent clear democratic oversight. Those events also gave governments renewed authority "to spy on their populations and invade recently won rights of privacy."[14] For Crouch these trends serve as another example of a postdemocratic turn, but there is something qualitatively different here that makes the authority—indeed, the advanced technological ability—to spy on the public more than merely another example of postdemocracy. Inasmuch as the authority to spy on the public has itself been privatized as part of the neoliberal policies of divestment, it marks a very specific evolution in postdemocracy as a political formation. In short, the proliferation of surveillance technologies in the hands of private enterprise is postdemocracy's final fortification.

As surveillance is woven deeper into the fabric of daily life, those same technologies add substantial weight to Crouch's claim that a return from postdemocracy back to a more egalitarian form of democracy is improbable if not altogether impossible—a claim that he makes independent of any real consideration of surveillance or its sociopolitical impact as a method of fortification. Enhanced by the increasingly subtle technologies of surveillance, postdemocracy is even more convincingly a point of no return: it is the beginning of a new era in political formations where, under the ostensible guise of democratic institutions and of the rhetoric of democracy itself, Western nations drift toward the "pragmatic" or "soft" authoritarianism of countries like Singapore and Malaysia and toward the oligarchical structures of post-Soviet Russia. "Egalitarians cannot reverse the arrival of post-democracy," Crouch maintains, "but we must learn to cope with it, softening, amending, sometime challenging it—rather than simply accepting it."[15] The question is how? The issue is not so much that surveillance has complicated the possibilities for softening, amending, or challenging postdemocracy but rather that surveillance arguably reconfigures postdemocracy into a more virulent form, fortifying its power, extending its scope, and reifying it as a political formation. The goal of surveillance is nothing short of a permanent change in the rules, and the rise of the surveillance society thus calls for a careful consideration not only of how its intrusive technologies reconfigure postdemocracy as a political formation, but—and this is the decisive issue for art practitioners—of what that reconfiguration ultimately demands of theater and performance.

Mr. Security, Surveillance, and the Vision of Public Space

For artists like Christoph Faulhaber and Lukasz Chrobok such reflections led to a series of direct confrontations with security forces outside of U.S. embassies and consulates in Germany and Poland over a two-year period from 2004 to 2006. The confrontations were part of a multifaceted project that they entitled *Mr. Security*, which they named after the fictitious private security firm that they created as the pretext for their actions within the highly surveilled areas around U.S. embassies and consulates. Outside of the embassies in Berlin and Warsaw and the consulates in Munich and Hamburg—as well as outside of a well-guarded jewelry store in Berlin, outside of the 2005 "International Conference on Security Policy" in Munich, and even in the customs and immigration area of the John F. Kennedy Airport in New York—Faulhaber and Chrobok donned the attire of private security personnel, positioned themselves conspicuously in locations that put them in direct competition with actual security officers (both private and public police forces), and thus deliberately provoked confrontations with officials who were both alarmed and bewildered not only by the inexplicable presence of Faulhaber and Chrobok but also by the bizarre combination of their seeming mastery and yet mildly tongue-in-cheek use of security rhetoric and protocol—all of which positioned Faulhaber and Chrobok within a carefully calculated aura of ambiguity where it was never quite clear whether they were or were not legitimate officers of an actual private security firm.

The confrontations were complicated further by the fact that Faulhaber and Chrobok matched the embassies' and consulates' use of surveillance cameras with photography and video cameras of their own. While these cameras were clearly part of the larger security accessories or props that Faulhaber and Chrobok used in their performance, they were also part of the functional apparatus that they used in their meticulous documentation of this initial phase of their project, documentation that in addition to photos, video, and audio recordings ultimately included subsequent legal correspondences to and from the police and local and federal authorities, and from representatives of the U.S. government. The documentation also included published transcriptions of their confrontations, which Faulhaber and Chrobok were careful to record and transcribe. All of this became material for a series of exhibits in prominent galleries and exhibitions like the KX Gallery in Hamburg, the Basis Exhibition Platform in Frankfurt, and, most notably, the Documenta 12 in Kassel. But these exhibits were far removed from the actual sites of confrontation and more importantly from a concept of art and performance that is inseparable from literal acts of contestation—a concept that I would place beneath the larger paradigm of "performance postdemocracy." So rather than relegating their project solely to the sanitized and arguably disarmed spaces of museums and galleries, Faulhaber and Chrobok entered a new phase of their *Mr. Security* project. They called this new phase *Roadshow 2006*, and it took them and all of the material that they had previously assembled for their exhibits back to the original sites of confrontation, that is, back to areas around the U.S. embassies and consulates.

The road show was part performance and part mobile exhibit in the back a delivery truck that Faulhaber and Chrobok converted into a gallery space. Picking up where their initial performances had left off, the two activist artists donned their security uniforms once again and traveled back to the embassies and consulates—this time with the delivery truck, which, now under the performative auspices of *Mr. Security*, they characterized as a "Rapid Mobile Office," a designation accentuated by a corresponding *Mr. Security* logo that Faulhaber and Chrobok attached to its side. The truck with its prominently displayed logo brought them once again into close proximity with the surveillance regimes around the embassies and consulates since the two artists deliberately parked the truck in the vicinity of where their earlier confrontations with embassy and consulate authorities had transpired. Explaining the concept behind *Roadshow 2006*, Chrobok suggested that it was not enough to assemble the *Mr. Security* material for conventional exhibits at galleries and museums in what amounted to a retrospective on the initial phases of the project. In this respect, the gallery and museum exhibits were merely means to an end: assembly points for material and for acquiring cultural legitimacy for the next phase of their project. Only when they had returned to the original sites of contestation with that material on display in the back of the truck, Chrobok argued, would the depth and complexity of the collusion between U.S. and German authorities in the monitoring and surveillance of citizens become evident.[16] Location thus became the crucial element in a calculated mix of artistic practice and daring radical activism.

While there was certainly a didactic aim in Faulhaber and Chrobok's concept of a mobile exhibit full of—among other things—legal documents and correspondences with German and U.S. authorities, the van itself functioned as a powerful mechanism for encroaching upon, contesting, and opening up public spaces that otherwise had been fortified by excessive surveillance technologies, legally ambiguous security protocols, as well as armed security forces and police—all of which not only seemed to function within their own postdemocratic bubble but also to expand, redefine, and co-opt public spaces in the process. Challenging that expansion was at the heart of

Mr. Security as a project. Indeed, when officers from the Bavarian branch of the State Security Office unexpectedly showed up at Faulhaber's home shortly after the initial phase of *Mr. Security* and made inquiries about his activities, he told them: "I'm working on some projects on the subject of the public space" and explained that "the starting point for *Mister Security* is the question of whether it is prohibited to take pictures of the American embassy." When the lead investigative officer answered by telling Faulhaber that "there is no such general prohibition in the relevant laws" and that he hadn't "done anything illegal," Faulhaber responded, "That means the officers outside of the embassies gave me false information."[17] So if he had not done anything illegal, why then were agents of the State Security Office paying him a visit? The explanation that Faulhaber received—and it was an explanation fraught with Kafkaesque overtones—was the vague and ominously threatening statement that "a few circles in Germany are worried."[18] Like the technologies of surveillance themselves, the officers' visit was designed to intimidate through an implied but never stated threat.

For their part, what the investigative officers received was an understanding of Faulhaber's project as the product of a politically motivated artist-provocateur who measured the success of his work according to the degree that he was able to provoke the agents of surveillance into action and hence out into the open where they would be subject to political scrutiny. In fact, this understanding was punctuated in what was arguably the most telling (and perhaps most hilarious) moment of the surprise visit from the officers of the State Security Office. Faulhaber transcribed the interview on paper as it took place since the officers forbade him to do an audio recording of their visit. But with his transcription, Faulhaber was nonetheless able to secure a record of the precise moment not only when the officers finally understood the dynamic behind his artistic practice but also when it dawned on them that they had been snared within it:

> SI [SUPERINTENDENT] ERDINGER: You provide a stimulus and know the automatic behaviour that it triggers. . . . And with this project you know in advance what will happen: you enter a security zone, take pictures, a guard comes, takes your details, prohibits you from taking photographs, and you're banned from the area.
>
> [. . .]

SI ERDINGER: And when the police arrive, it's part of your project. That means you're working right now while we're sitting here?

FAULHABER: So are you!

SI ERDINGER: In other words, we're part of your project? That's why you wanted to record us? How do you want to present all of this? A catalogue? An exhibition?

FAULHABER: No, I don't know yet. I do many projects where I don't know beforehand how they're going to end.

CS [CHIEF SUPERINTENDENT] KRAMER: And you studied art?

FAULHABER: Yes, in Hamburg.

In fact, the transcription of this visit became part of the exhibits not only at museums and galleries but also within the Rapid Mobile Office with which Faulhaber and Chrobok returned to the scenes of the initial phase of *Mr. Security*. Once again, the goal was to provoke confrontation, but this time Faulhaber and Chrobok's work became a magnet not only for the embassy and consulate security forces but also for a curious public that, in small flash-mob fashion, began to congregate around the free mobile exhibit and thus implicitly to contest and to reclaim the areas around the embassies and consulates as free public cultural spaces.

At a conceptual level, there was nothing serendipitous about this effect. The threat that surveillance poses to the very notion of the agora was, in fact, the central issue that the publisher Oliver Zybok highlighted in the editorial preface he wrote to the small publication that had accompanied the earlier museum and gallery exhibits of the *Mr. Security* material—a publication that is itself a curious hybrid of an actual museum catalog, a transcription of Faulhaber's exchanges with the police and embassy and consulate security personnel, and a faux business brochure for the fictive private security firm *Mr. Security*. In that opening editorial, Zybok introduced readers to *Mr. Security* by highlighting the profound effect that surveillance technologies have had on our understanding of space, an effect that we might reasonably link to the ideological underpinnings of the neoliberal policies of divestment, outsourcing, and the transfer of public services into private hands. Indeed, echoing the notion that surveillance is but the fortification of a larger postdemocratic phenomenon, Zybok contends that "new privately operated spaces, dominated by cameras, security staff and rolling grilles . . . destroy the illusion that this space might belong to the community." Behind the surveillance cameras and security protocols, "There is always a power and a property relationship." Ultimately, "It is not uncontrolled and allegedly unoccupied space that we are losing," Zybok concludes, "but public space as a vision."[19] That lost vision of a public space is a vision grounded in the egalitarian assumptions of a democratic society and of the agora. It is a vision supplanted by the postdemocratic logic of the surveillance lens.

It is amid the fog of that lost vision that the performances of *Mr. Security* and *Roadshow 2006* conduct their social work. Not coincidentally, the catalog, for which Zybok wrote the prefatory editorial, ends with the image of a prototype of Faulhaber and Chrobok's delivery truck along with the assurance: "*Mister Security* will face up to the challenge of the increasing mobility and asymmetry of the global threat. One example of innovation and a new flexibility is the development of the Rapid Mobile Office (RMO©). Our goal will continue to be: To Serve and Observe."[20] The irony of this proclamation was quick in coming since, once set into motion, Faulhaber and Chrobok's "Rapid Mobile Office" was itself an asymmetrical weapon of cultural and political warfare—and a significant enough one, I think, to have served if not as the inspiration for, then at the very least as a precursor to activist artist Clark Stoeckley's "Wikileaks Truck," which became an important fixture in New York's Occupy Movement and which, among other things, offered shelter to protesters. Although Stoeckley's truck had no real connection to Wikileaks, police at one point actually impounded it based on the unlikely assumption that Stoeckley was using it to distribute classified material to protesters and hence posed a national security threat.

But there was something more tactical about the Rapid Mobile Office. Whether Faulhaber and Chrobok, in envisioning *Roadshow 2006*, planned for a collective contestation of embassy and consulate perimeters with the help of the public is not altogether clear—although they did envision that the spectacle of their own performance, the odd presence of their Rapid Mobile Office, and the open invitation to the public to examine the van's contents would capture the interest of passersby. That all of this might result in a crowd of people congregating along the perimeters of U.S. embassies and consulates in a seeming impromptu demonstration and critique of one of the most prominent manifestations of the international surveillance industrial complex was no small matter. Nor was the presence of an unidentified delivery truck at the edge of that perimeter. But the same had been true earlier of Faulhaber and Chrobok's appearance in the initial phases of *Mr. Security* when they arrived at the embassies and consulates in private security uniforms and started taking photographs and video footage. Returning to those initial confrontations momentarily, I want to note that, despite all of the surveillance technology and CCTV cameras employed by the embassies and consulates themselves, it was the cameras used by Faulhaber and Chrobok that ironically proved to be the most contentious element in their performance, as the embassy and consulate guards were consistently alarmed less by the physical presence of Faulhaber and Chrobok as unidentified security personnel than by the fact that they were taking pictures and video footage of the area surrounding the embassies and consulates as well as of the official guards themselves. The guards' concern was that people they did not know were taking photographs and footage for purposes that were uncertain and that this material, in turn, might be further disseminated in ways that could not be identified or controlled and that it thus might be used to the detriment of all. In short, Faulhaber and Chrobok effectively transformed the embassy and consulate guards into the spokesmen for the critique that *Mr. Security* offered of surveillance more generally.

At the most immediate level, the logic behind that alarm was the obvious fear that Faulhaber and Chrobok might be preparing or helping others to prepare for some unspecified future threat that the collection and dissemination of photos and video footage to persons unknown might facilitate. As one of the police officers at Hamburg's Consulate General of the United States told Faulhaber as he demanded that Faulhaber stop taking photographs: "The Americans feel threatened."[21] It is hard, or course, to miss the irony of officials engaged in large-scale surveillance of their own feeling threatened by an artist taking photographs. But if the irony here were not rich enough, the only legal ordinance that police officers ultimately could cite in their demands that Faulhaber cease using his camera was the Kunsturhebergesetz, that is, the Artworks Copyright Law, which the chief superintendent at the U.S. embassy in Berlin told Faulhaber he had "probably violated . . . by taking pictures of . . . [his police colleagues] without getting their permission first." It was bad enough that in their efforts to thwart Faulhaber's performance art, the police officers hid behind a law designed to protect artists and their intellectual property, but the superintendent actually misrepresented the particulars of the law that he cited. While section 22 of the Artworks Copyright Law does explicitly state that "depictions of a person or persons may not be distributed or publicly displayed without the consent of the person or persons depicted," exceptions to this restriction exist when "the distribution or public display of . . . [such depictions] serves a higher artistic purpose, provided that they have not been commissioned."[22] Once again, Faulhaber and Chrobok effectively transformed the embassy and consulate guards into the spokesmen for their own work and cause. This time they inadvertently provided Faulhaber and Chrobok with the legal foundation for their project as artists while simultaneously raising questions about the massive collection and dissemination of photographs and video footage by the private security forces employed at embassies and consulates who, while having access to state-of-the-art surveillance technology, were unlike Faulhaber and Chrobok in that they had no identifiable "higher artist purpose."

That higher purpose did not emerge within the safe space of a studio, gallery, rehearsal room, black box, or conventional theater. It unfolded on the highly surveilled perimeters of a security zone. When all was said and done, *Mr. Security* had maintained a delicate balance not just between art and politics. It created situations that were equal part playful, unpredictable, and absurd on the one hand, and yet tense, volatile, and genuinely dangerous on the other. However humorous these situations might appear in retrospect or in carefully edited film, there was always something deadly serious about them. Faulhaber and Chrobok were in fact provoking armed guards who not only had a broad mandate in their security assignments but who could themselves be unpredictable in situations that were perceived to be peculiar and hence potentially threatening. In this respect, there is a significant difference in the stakes of Faulhaber and Chrobok's *Mr. Security*, and the politically provocative hoaxes orchestrated, for example, by activist-artist groups like the Yes Men. Critics like Mark Dery and Marilyn Frankenstein have rightly celebrated the Yes Men's political aesthetic of "culture-jamming," that is, their mastery of "the art of interrupting and disrupting prevailing cultural signals and codes."[23] This the Yes Men have done by successfully impersonating representatives of corporations like Dow Chemical or of organizations like the World Trade Organization. Then, by making statements in the media that prove to be awkward for those they have impersonated, the Yes Men attempt to nudge the actual corporations and organizations in politically progressive directions.

Such antics, along with the documentary films made about them, have gained the Yes Men a well-deserved reputation as politically progressive pranksters, or as what Dery has called "Groucho Marxists,"[24] and I do not want to undervalue the significance of the creative activism they practice. But there is a degree of what Richard Schechner has called "dark play"[25] in Faulhaber and Chrobok's *Mr. Security* that sets them apart from the Yes Men, that is, an element of genuine risk and danger in their political-aesthetic, especially since it was calculated to provoke interaction with forces capable of dangerously unpredictable responses. We might measure the level of that risk not just by the severity of the responses their project triggered but also by the kind of forces it provoked into play and brought out into the open. Their work involved the kind of dark play that in 2005 resulted in those officers from the State Security Office showing up unannounced at Faulhaber's Munich home or that later that same year unexpectedly landed Faulhaber in an overnight stay in a detention cell at the border between Canada and the United States, where, as he traveled from Montreal and New York on a Greyhound bus, immigration officers forcibly detained and interrogated him, much to his surprise, about his financial resources and "about the project . . . [he] had been pursuing with the surveillance of the US Embassy in Berlin."[26] In both instances, the purpose was not law enforcement but state-sponsored intimidation.

Inasmuch as those moments were actually triggered by the initial phases of *Mr. Security*, they too were part of the larger project and its politically charged entry into the unpredictable realms of dark play. In this respect, the *Mr. Security* project arguably continued to reverberate for years to come in both positive and negative ways. On the positive side, recognition of the political and artistic significance of the project led to Faulhaber being awarded a stipend from the Cultural Foundation of the State of Rheinland-Pfalz in support of his having also been awarded a six-month resident fellowship at New York's Location One—a fellowship that was itself a recognition of the artistic significance of *Mr. Security*. But on the negative side, the fellowship was complicated by seemingly inexplicable delays in the processing of Faulhaber's visa, delays that blocked his planned residency by over a year, and even afterward, when Faulhaber finally arrived in New York, he was immediately met and detained again as he exited the airplane, this time by Homeland Security officers. Shortly thereafter, the administrative board at Location One informed him that his residency had been terminated. The reason for his termination, according Location One's executive director, was that "certain of . . . [Faulhaber's] actions" had come "to the attention of United States national security authorities"—ironically, these were the very actions that were part of the *Mr. Security* project, which had won him the support that he had received from Cultural Foundation of the State of Rheinland-Pfalz and from Location One. The executive director explained further in diplomatic but ominous tones that "after consultations with law enforcement agencies and . . . [with their] own legal counsel," they had "determined that the risk to Location One" if Faulhaber continued in their "program" would be "too great for" for them "to accept," and so they "asked him to leave."[27] The subtext to the termination was a lesson in the immense flexibility and power of surveillance regimes more generally: if Faulhaber could not be intimidated himself, those around him could.

Performing "A Bit of Damage," or Postdemocracy's Inevitable Products

At some level, such moments are little more than minor skirmishes in the larger context of what, under the general rubric of "performance postdemocracy," we might identify as an emergent strategy of cultural warfare that combines creative activism and direct confrontation and that sees this combination as one of the few remaining options for drawing nebulous security forces and surveillance regimes out into the open where their extraordinary legal status might be subjected to scrutiny and where the discrepancies between the rhetoric of democracy and the realities of postdemocracy might be challenged. Whether such challenges are the rehearsal for new more egalitarian political formations is an open question. But a project like *Mr. Security* points toward a theatrical practice that looks well beyond Crouch's resigned prescription for "softening" or "amending" the political realities of postdemocracy or the concomitant realities of the surveillance society. In this respect, Faulhaber and Chrobok's work foreshadows the more militant role for theater and performance that Alain Badiou has advocated in his book *Rhapsody for the Theatre*.

Although stated in very different terms than those laid out by Crouch, Badiou's own assessment of the contemporary political landscape is equally dire. For Badiou, that landscape is nothing short of an "inegalitarian monstrosity," which is itself the offspring of "liberal capitalism" and a "parliamentary system" of democracy so plagued by "insidious corruption" that it is little more than a theatrical sham maintained for what Badiou describes as continued "oppression in the form of a rigged choice."[28] In brief, Badiou's critique is of an insular and self-serving political order that functions for the few at the expense of the many and that sustains itself in the masquerade of democracy's troupes and in the fortifications of the surveillance society. But if there is an aura of Crouch's notion of postdemocracy in Badiou's diagnosis, what sets the two apart is Badiou's sense of what current political realities demand. Badiou chooses a rougher terrain than Crouch's path of resigned amelioration. Suggesting that we must "use all the means at our disposal to begin to undermine the reactionary consensus,"[29] he argues that the pursuit of "true justice is certainly worth a bit of damage on the side of those who are well-off" and on the side of "their servants."[30] This is a prescription emanating from a basic loss of faith in the continued viability of the democratic process to correct social and political injustice. It is a militant prescription in response to postdemocratic realities, and it calls for a very different kind of theater than that which works comfortably within the existing institutional structures of Western democracy that we have known and embraced heretofore. Indeed, the value that Badiou places on perpetrating "a bit of damage" in order to effect genuine political change implicitly calls for a new militancy in theatrical practice as well. Desperate times produce desperate measures.

One can certainly debate the merits of a theater that does not shy away from doing damage. But we must also ask where the call for such a theater originates in the first place. In many respects, this is the more important question, and though Badiou has a long history of embracing radical politics, the answer to this latter question amounts to more than a bit of sixties revolutionary nostalgia. So before rejecting Badiou's call as a nostalgic, naive, or even dangerous political rationalization of the spectacle of vandalism or of a theater bent on acts of destruction and damage in the streets, we might consider his implicit call for a new militancy in theatrical practice as an inevitable product of the forces of postdemocracy themselves—a product that increases in intensity as those forces become more fortified, entrenched, and global, and that is also exacerbated by the proliferation of surveillance regimes. The logic behind such a suggestion is not altogether new. At one level, it harkens back to my earlier modification of Marx's argument from *The Eighteenth Brumaire* and to the notion that artists do not make theater "under circumstances chosen by themselves, but under circumstances directly encountered, [and] given." But on a more contemporary note, it echoes the logic that is central to Jean Baudrillard's book *The Spirit of Terrorism*, which, while written in response to the events of September 11, 2001, has as much relevance today as it did when it was first published in 2002.

That relevance centers on the dynamic that Baudrillard identifies behind the events of September 11. He argues that those events—and the damage that they did—were, in fact, the inevitable product of the hegemonic processes of globalization, processes that scholars like Taitu Heron have argued are fueled by neoliberal ideologies[31] and that Baudrillard, similar to Crouch, maintains have steadily but effectively undermined Western democracy and led to what, in all but name, he describes as a postdemocratic era. But in contrast to Crouch's arguments, there is no resigned acceptance at the end of Baudrillard's analysis. On the global stage, the processes of postdemocracy reap what they sow, producing reactions not of resignation and submission but of deep resentment and tactical retaliation—the kind of retaliation that, as events like September 11 remind us, may be asymmetrical in tactical structure, but are potentially lethal in effect. So while Baudrillard and Crouch may agree in linking the unfettered rise of global capital with democracy's demise,[32] Baudrillard's argues that the processes of global capitalism, which are spearheaded by neoliberal ideologies, are tantamount to a form of terror, and economic terrorism, he argues, is dialectical in effect. In response to the economic havoc it creates, one of globalization's inevitable returns on its investments is thus a variation of the ruthless aggression with which it cultivates and fortifies its markets. "So it is terror against terror," Baudrillard argues. "But asymmetrical terror."[33] Whereas Crouch argues that the policies of neoliberalism have ushered in an era of postdemocracy with which we simply must contend, Baudrillard argues that the related processes of global capitalism have changed the rules on all sides and have done so in such a way that extraparliamentary action (and in particular violent extraparliamentary action like the attack on the Twin Towers) is the predictable recourse. Indeed, such acts are globalization's inevitable product because, so Baudrillard reasons, they are the only option left. "When global power monopolizes," he argues, "when there is such formidable condensation of all functions in the technocratic machinery, and when no alternative form of thinking is allowed, what other way is there but a *terroristic situational transfer*?"

Answering his own rhetorical question, Baudrillard concludes: "It was the system itself which created the objective conditions for this brutal retaliation [on September 11]. By seizing all the cards for itself, it [i.e., global capitalism] forced the Other to change the rules. And the new rules are fierce ones, because the stakes are fierce."[34] These new rules, I would argue, play out not only on the global stage but on the domestic one as well. For if attacks like those of September 11 and the ones that have followed are globalization's inevitable product, so too are the technologies of security and surveillance that are deployed in service of those who calculate the threat of terrorism as a cost of conducting business and who, as a consequence, also advocate increasingly expansive and ever more intrusive security and surveillance protocols both at home and abroad in order to protect the market for their international business ventures. Indeed, at one level, private enterprises have a vested interest in cultivating and sustaining the aura of threat as a strategy for maximizing and securing their own further profits. But is that aura of threat really the only possible antinomy produced the processes of globalization?

As compelling as I personally find Baudrillard's arguments—even a decade and a half after they were formulated—the notion of a "terroristic situational transfer," which Baudrillard posits as the single possible reaction cultivated by the economic violence of globalization, is too narrowly defined. Indeed, deep within Baudrillard's notion of an asymmetrical terror, which is itself the inevitable product of economic terrorism (both global and local), we might discover not merely the seeds for acts of horrific violence like those of September 11, but alternatively the seeds of an unavoidable, indeed necessary, dynamic in the theater and performance of postdemocratic surveillance societies. We might discover the seeds for something structurally akin to "terror against terror" and something that in its asymmetrical structure accounts for Badiou's implicit call for a new militancy in theatrical practice. In short, we might discover the outlines of a theatrical practice that, while not aligned with heinous and murderous acts like those perpetrated on September 11, 2001, is nonetheless committed to doing "a bit of damage" in order to challenge postdemocratic formations and press for social justice. Buried deep between the lines of *Rhapsody for the Theatre*, I would thus argue, one might find an understanding that asymmetrical reactions to entrenched and fortified antidemocratic formations are by definition much richer, more diverse, and inherently more creative in their potential than Baudrillard allows.

It is in this diverse, creative asymmetry that Badiou's call reverberates and that the notion of militant theatrical practice as a vibrant antinomy to postdemocratic surveillance societies can be found. So rather than "terror against terror," or even asymmetrical terror against terror, the kind of theater and performances necessitated by postdemocratic surveillance societies, I would argue, are those that engage in asymmetrical and creative moments of disruption; those that engage in tactical acts of damage as activist art; and those that unfold in environments where the rules and the stakes are, as Baudrillard notes, "fierce." To some extent, the basis for this argument is already present in the periphery of Baudrillard's own book, where stepping back momentarily from the particulars of terrorism itself, he argues that "at the very heart of the process[es] of globalization" is "the emergence of a radical antagonism" of which terrorism as a response is only one example among different possible responses.[35]

Since the model that Baudrillard suggests in his formulation, "terror against terror. . . . but asymmetrical terror," is one of disproportionate power relations, it would stand to reason that the antagonism, however radical it might be, must be guerrilla-like in its tactics. In fact, this appears to be what Baudrillard intends early in his book when he describes terrorism as "a new form of action which plays the game, and lays hold of the rules of the game, solely with the aim of disrupting it."[36] Extending this same logic further, I want to suggest that the political realities of postdemocracy and the surveillance society not only necessitate a theatrical practice that assumes acts of disruption and civil disobedience to be art's point of departure but that also, rather than envisioning these acts as ends in themselves, sees them instead as a necessary means for a radical transformation of the public sphere, even if such transformations occur one act at a time.

From Virtual Disturbance to Physical Damage

This suggestion obviously echoes the kind of theatrical practice that groups like the Critical Art Ensemble and the Electronic Disturbance Theater initiated in the 1990s just as Internet technologies were coming of age. Even the conceptual contexts for this suggestion resonate with that echo. Looking back at the Critical Art Ensemble's classic 1994 essay "Nomadic Power and Cultural Resistance," for example, one would have difficulty missing the numerous ways that it foreshadows Baudrillard's later arguments about globalization and terrorism. Though the Critical Art Ensemble never identifies the forces of globalization by name—choosing instead the neologism "nomadic power"—it is clear that their notion of "nomadic power" is related to the postdemocratic political order that is at the core of globalization's processes and that is the focus of Baudrillard's arguments. Not only are twentieth-century (and twenty-first-century) capitalist manifestations of "nomadic power" "predatory" in nature, transnational in structure, and tailored to "the power elite of late capital,"[37] they "can no longer be disrupted," the Critical Art Ensemble argues, "by strategies predicated upon the contestation of sedentary forces,"[38] which is to say, by the established, governing institutional structures of state democracies. These descriptions mesh well with Baudrillard's later characterization of the processes of globalization, and almost a decade before he wrote *In the Spirit of Terrorism*, the Critical Art Ensemble had already argued that rather than succumbing to "acquiescence and complicity" with nomadic power's postdemocratic tendencies, artists and cultural activists should produce "act[s] of disturbance" and disruption, an admonishment that they justified, much as Baudrillard was to do later, with the simple rhetorical question, "What other chance is there?"[39]

In that last decade of the twentieth century, groups like the Critical Art Ensemble saw the greatest chance for viable creative acts of disturbance and disruption in a mastery of the rules and protocols of Internet technologies, which, as early hacker communities like the Cult of the Dead Cow, the Ninja Strike Force, Hacktivismo, and, later, Anonymous all skillfully demonstrated, lent itself well to asymmetrical activism. There was, of course a strong dose of playing and laying "hold of the rules of the game, solely with the aim of disrupting it"[40] in hacker communities, which they simply described as hacking for the "lolz." But the Critical Art Ensemble was among the first to recognize something profoundly performative in this new potential arena for creative activism, not only with regard to the functioning structures of the Internet itself but also with regard to how the Internet—which like nomadic power is literally without a national or urban base—was then and remains today a powerful and influential microcosm of late capitalist society.

With this recognition, the Critical Art Ensemble sought an activist theatrical practice that could skirt the fence between the virtual and material worlds. Indeed, Brian Holmes recounts that the early Critical Art Ensemble "dreamt of a recombinant theater, where the performative matrix would stretch . . . to the virtual world of the database" and where theater itself would be "modeled on the exploits of a hacker, using coding techniques to dismantle repressive data bodies."[41] Even at this early stage of the Internet's development, there was in the oppositional structures of the Critical Art Ensemble's dream a prescient recognition of the need to conceptualize and embrace a politically viable hacker aesthetic and thus to position artistic practice as an antinomy to data collection, dataveillance, and ultimately surveillance more generally. For if, in this dream, the image of the hacker—which is in fact an image of someone willing to do a "bit of damage"—passed seamlessly from the digital realms of the virtual to the concrete realms of material theatrical practice, this is because even in the nineties, groups like the Critical Art Ensemble recognized that the digital was fundamentally reconfiguring the political realities of the material world and providing the technological infrastructure for and the fortification of a postdemocratic order.

The Critical Art Ensemble's prescience notwithstanding, much over the past two decades has changed in the technological landscape that exceeds their premonitions. Not only is the Internet no longer "a raw, anarchic, electronic frontier."[42] It has evolved into a commercialized tool that, under the guise and rhetoric of providing new freedoms, has in fact lured vast swaths of the population into a sophisticated surveillance network designed to harvest and process mass data as marketable products for corporations like Microsoft, Google, Apple, Amazon, and Facebook. At one level that evolution is a profound affirmation of Jon McKenzie's 1999 claim that "there is no human-computer interaction that does not presuppose complex social and technological networks, and with them the values and power setups they enable and are enabled by."[43] Twenty years later, it is increasingly difficult to find spaces devoid of some form of human-computer interaction, and I would suggest that McKenzie's claim is outdated only in terms of its limited scope. The once seemingly clear distinctions between the material and virtual have blurred into what in popular cultural discourse is commonly referred to as "the Internet of things," and so McKenzie's claim now touches upon almost all facets of Western society.

Here we arrive in the full loop of this book, finding ourselves at its conclusion on those same smart streets where the book began. As we may recall from the discussions in the book's introduction, "the Internet of things" simply refers to a world where the proliferation of embedded technology and digital devices in everyday objects on the body (cell phones, wearables, health gadgets, etc.), in the home (thermostats, kitchen appliances, TVs, laptops, tablets, etc.), and in the public sphere (store entrances and checkouts, traffic intersections, public transportation systems, ATM machines, etc.) is so widespread that, to borrow a metaphor from Keller Easterling, "the computer has escaped the box."[44] The Internet of things refers to a world where "ordinary objects are carriers of digital signals,"[45] where, as a result, the physical spaces of the home and of the city have become an unavoidable network wired for the production and collection of data, and where we spend less time communicating with the smart objects around us than they do communicating with each other about us and collecting those communications for large data banks. In this respect, "The internet of things is not a revolution in the way the Internet once was," at least in the sense of that much-celebrated mirage called "net neutrality." It is rather what Bruce Sterling describes as a "power grab."[46] Our relationship "to the Internet of Things is not democratic," he argues. "It's not even capitalistic. It's a new thing. It's digital-feudalism."[47] The Internet of things is, in short, a euphemism that camouflages the advent of a new generation of surveillance technologies. It is, to echo Sterling, an "all-purpose electronic automation through digital surveillance by wireless broadband."[48] It is the technological fortification of postdemocratic society. But as the early work of the Critical Art Ensemble reminds us, this fortification comes about because material realities were already trending toward postdemocratic formations in the first place.

In retrospect, the Critical Art Ensemble's dream of an activist theater "modeled on the exploits of a hacker" was a reaction, perhaps more than anything else, against the emergent forms of postdemocracy that even then were as pronounced in the digital structures of virtual reality as they were in the tangible structures of the material world. The intersection of these two forms is even more pronounced today, at least according to cultural critics like Kruk and van der Velden. In their recent book, *Black Transparency*, for example, they argue: "The working model of self-government that we do have—parliamentary representative politics—is, at best, an ailing patient,"[49] and they maintain that "every person's capacity to evade surveillance is determined by their position in a feudal matrix of technological and institutional dependencies."[50] In this now familiar allusion to the decline of democratic institutions and the rise of technologically enhanced postdemocratic formations, Kruk and van der Velden find the basis for what is arguably the most provocative rhetorical question in their book: "Isn't a structural lack of democratic governance over the internet somehow mirroring the democratic deficit—the fundamental shortcomings and boredoms of parliamentary democracy?"[51] Provocative though this question may be, the image of "mirroring" that Kruk and van der Velden summon in their description is a poor metaphor for what might more accurately be described as an alarming reciprocal dynamic in which the "lack of democratic governance over the internet" and the failings of "parliamentary representative politics" have fed each other's postdemocratic proclivities and cultivated the growth of an advanced surveillance society. Neither can be viewed in isolation.

The question is how they might be resisted. Back in 1999, McKenzie argued that with the advent of digital technologies, resistance needs to take the form of what he called "interhacktivity," or hacking that, while focused on "the interactivity between humans and computers . . . not only takes aim at technical systems, *but also targets social systems*"[52]—which brings us back to the early Critical Art Ensemble. Like McKenzie they too saw a clear path from digital to material political realities. Indeed, if the political dynamics of the Internet provided a microcosmic window into the political dynamics of Western society more generally, then the Critical Art Ensemble's interest in hacker communities becomes all the more understandable. The potential political significance of those communities resonated well beyond the nebulous boundaries of the virtual world and suggested a model of opposition for the sites where the more repressive currents of the virtual and the material worlds collude—sites where today the operational structures of the surveillance society are present and active and wield immense authority and power, sites whose numbers increase unabated.

In practical terms, the Critical Art Ensemble's interest in the image of the hacker suggests that they looked to hacktivist disruptions of digital technologies in order to find performative models of disruption that, in turn, could be circuited back and extended to creative activism against the postdemocratic structures of surveillance societies more generally. In this respect, McKenzie's claim—"interhacktivity, hacking, interactivity: all must be understood as effects of performative power—and as its instruments"[53]—echoed the political-aesthetic sensibilities of not only artist-activist groups like the Critical Art Ensemble but also offshoots like the Electronic Disturbance Theater. Focusing on the work of this latter group, Catherine Bernard formulated similar arguments just as the new millennium was beginning, and her arguments are worthy of note if only because she formulated them in conceptual terms that were significantly more refined than Baudrillard's subsequent argument that the economic terrorism of global capital inevitably produces a terrorist antinomy. There is no talk of "terroristic situational transfers" in Bernard's work but of vulnerabilities that might be exploited—vulnerabilities that in Internet technologies hackers take advantage of and that, by way of their example, open the door to the models of what the Electronic Disturbance Theater embraced as a radical theatrical practice and to what throughout this chapter I have called "performance postdemocracy."

Bernard's specific concern was with artistic counters to the commercial structure of cyberspace by groups like Floating Point Unit

Fakeshop and, more significantly, the Electronic Disturbance Theater. In her 2000 article "Bodies and Digital Utopia," she argued, for example, that through their own unique approach to artistic activism "the Electronic Disturbance Theater [has] deftly demonstrate[d] that cyberspace contains, within its structure, resistance tools" that might counter the "politics of repression."[54] The larger argument of this important essay was an implicit reminder that power inevitably produces its own vulnerability and thus also the tools with which to resist repressive politics as well. The same is true, I would argue, with those sites where the more repressive currents of the virtual and the material worlds collude. How activists and artists exploit such vulnerabilities—how they "hack" them—is a matter of strategy and choice, which in turn is also a matter of means and ends or goals. But the tools, Bernard argues, are there. It falls only to those bold and creative enough to take them up, to risk "a bit of damage," and to disrupt the rules of repression's game.

Of DDOS, Firewalls, and *Threat*: *Lubyanka's Burning Door*

At some level, it is unfortunate that Bernard did not explicitly broaden the scope of her discussions beyond the political implications of the work of the Electronic Disturbance Theater (and Floating Point Unit Fakeshop) to include other arts practitioners who, in their own right, deserve to be counted among the radical antagonists produced by disproportionate power relations and who, while having strong conceptual affinities with hacker communities—at least, in the sense of laying "hold of the rules of the game, solely with the aim of disrupting it"—have exploited power's self-produced vulnerabilities using very different tools of resistance than those literally aligned with hactivism and "electronic civil disobedience."[55] But rather than leaving this statement as an implicit wish that Bernard had somehow written a different article, perhaps the more constructive approach is to take this wish as an opportunity to update her arguments and thus to close this chapter and this book with a comparison that uses what is widely considered to be the Electronic Disturbance Theater's most significant act of creative activism from the late 1990s—namely, the troupe's innovative support of the Zapatistas' rebellion in Chiapas,

Mexico—as a conceptual frame for understanding the significance of the work of the Russian performance artist Pyotr Pavlensky and in particular for understanding his piece *Threat: Lubyanka's Burning Door*, which began on November 9, 2015, when he doused the front doors of the Russian Federal Security Service building (the former KGB building) with petrol, set them on fire, and then "waited calmly, with his back to the flames, holding the gas can in his hands as the authorities descended"[56] and arrested him—a lone figure stoically waiting just outside Lubyanka's literal firewall, a lone figure making no attempt at entry but making his presence known. Here was a political artist willing to do a bit of damage.

My own rhetorical rendering of that moment consciously bridges it to what is arguably the Electronic Disturbance Theater's most important work of creative activism, their development and deployment of FloodNet in the late 1990s on behalf of the Zapatistas' cause. Indeed, it was the Zapatistas' rebellion in Chiapas that "provided the impetus for the formation of the Electronic Disturbance Theater"[57] in the first place. As is well known, FloodNet is a "distributed denial of service"—or DDOS—program that operates by "exploiting the Java applet reload function"[58] on websites. It is thus capable of flooding targeted servers with access requests so numerous that it disrupts the server, which cannot process the numerous requests quickly enough and which, as a consequence, either temporarily crashes or slows to a crawl because it is literally overwhelmed with traffic. In 1998, the Electronic Disturbance Theater "held several pro-Zapatista actions . . . using FloodNet . . . [to target] websites ranging from those of the Clinton White House and the Pentagon to those of Mexican president Ernesto Zedillo and the Frankfurt Stock Exchange."[59] As Jill Lane notes in her article "Digital Zapatistas," "No data was destroyed" and "No webpage was altered." Metaphorically speaking, there was merely an activist mob amassing outside of the targeted servers' firewalls and knocking at its entrances in virtual protest. In "one iteration," Lane observes, "the FloodNet repeatedly requested nonexistent pages" from the site of the Mexican government. These requests used names such "as 'justice' or 'human rights'" and thus compelled "the server to produce a steady, flashing stream of '404 error-reply' messages stating: 'justice not found on this site' and 'human rights not found on this site.'"[60] Referring to similar acts of electronic civil disobedience in her recent book *Hacker, Hoaxer, Whistle-*

blower, Spy: The Many Faces of Anonymous, Gabrielle Coleman credits the Electronic Disturbance Theater with pioneering a unique brand of "virtual sit-ins" that "combined technical interventions with poeticism and performance art."[61] This combination was in no small part the brainchild of Ricardo Dominguez, who originally was a member of the Critical Art Ensemble before founding the Electronic Disturbance Theater with Stefan Wray, Carmin Karasic, and Brett Stalbaum. It was Dominguez who specifically sought to translate the "social aesthetics" of figures like Bertolt Brecht, Augusto Boal, and Luis Valdez into a vision for the "digital stage."[62] But if I may, I would like to flip the circuit of Dominguez's vision and return us back to the streets twenty years later. As I have argued throughout this book, with the advent of advanced surveillance regimes and smart technologies, the streets are now every bit as much a part of the digital stage as they are a part of the material one. It is on those streets that we find Pyotr Pavlensky . . . at least temporarily.

That *Lubyanka's Burning Door* begins on the streets is easier to establish than determining where precisely the piece ends, for it was calculated as an intervention that extended by design not into virtual spaces but into a dizzying legal maze as Pavlensky effectively laid hold of the arcane rules of the legal game, making a mockery of the charges brought against him, of the trial where he was prosecuted, and of the conviction it ultimately yielded. The courtroom is always a venue of performance, but Pavlensky successfully transformed it into an extension of the politically radical performance art that he began on the street with *Lubyanka's Burning Door* and continued throughout the proceedings against him. Like the Electronic Disturbance Theater before him, which linked its artistically inspired digital activism to the Zapatistas' revolt in Chiapas, Pavlensky linked his performance art activism to the cause of anti-Russian activists and artists in the Ukraine and later also to the cause of the Primorye Guerrillas, or, as they are often called, the "Primorsky Partisans," who were arrested after taking up arms in a revolt against police brutality and corruption "in Russia's Primorye, or Maritime, region near the Chinese border."[63] In both instances, Pavlensky allied his work with artists and activists who were willing to do a bit of "damage" in principled acts of resistance to police terrorism. This became evident as soon as the prosecutor filed charges against Pavlensky after his arrest.

Initially, Pavlensky was charged with vandalism, but he surprised the court by answering the charges with the demand that they be changed to "terrorism." In part, this was a show of support for the Ukrainian film director Oleg Sentsov, who in 2014 had burned down "the door to the Crimean office of the pro-Putin United Russia party"[64] and who, as a result, had been "sentenced to 20 years [in jail] on terrorism charges."[65] But Pavlensky's demand was also a strategy that highlighted his own act as something akin to Baudrillard's notion of asymmetrical "terror against terror." As Anna Nemtsova reported in the American press just two days after Pavlensky's arrest: "Before setting the security headquarters on fire, Pavlensky wrote . . . [that] 'the burning of Lubyanka's door is a glove [i.e., a gauntlet] that our society throws into the terrorist threat introduced by the FSB [Federal Security Service], which is using its methods of endless terror to keep power over 146 million people.'"[66]

What is noteworthy in this public statement, particularly when one compares it with the subsequent statements that Pavlensky made upon his release, is the general lack of distinction that he makes between the notions of terrorism and surveillance, the latter of which, while unmentioned here, figures prominently in subsequent statements. I will discuss those latter statements momentarily, but for now suffice it to note that Pavlensky clearly understands the state's use of surveillance as a key mechanism in its ability to terrorize citizens, and in his statements after the trial, Pavlensky is precise enough in this regard that one is left with the sense that his earlier demand to be charged with terrorism had as much to do with his use of the basic tools of surveillance like the video and photo cameras that documented the initial event (including his arrest) as it did with vandalism itself. Inasmuch as these are also the tools that the FSB uses to terrorize the Russian population, they were also a part of the artwork's antisurveillance aesthetic, which involved a conceptual high-jacking—a "hacking" as it were—of the language, tools, and institutional structures of state-sponsored surveillance.

While the use of those cameras as well as the media spectacle that Pavlensky created in the courtroom was as much a part of the performance as the actual burning of Lubyanka's doors, the prosecutors were loath to meet Pavlensky's demand. But they did change the charges. With no identifiable sense of irony or of the way that Pavlensky was playing them into his own performance, the Russian prosecutors countered not by charging him with terrorism, but—in a bizarre coincidental parallel to German police telling Christoph Faulhaber that he had violated the Artworks Copyright Law—by shifting the charges against Pavlensky from "ideologically motivated vandalism" to "damaging a cultural heritage site."[67] In both instances (Faulhaber's and Pavlensky's), antisurveillance interventions exposed a tendency within the surveillance industrial complex to invoke culture as a last line of defense.

In Pavlensky's case, this new charge was based on a designation that the former KGB building had received in 2007 and that prosecutors literally justified by stating publicly that the FSB building had to be protected because so many "leading figures of science and culture" had been "imprisoned there"[68] and "hundreds of 'outstanding cultural figures'" had been "tortured in its cells."[69] Not only did Pavlensky respond by telling the press that those working in the FSB building had "been methodically destroying . . . [Russian] culture for nearly 100 years now, while at the same time having the gall to publicly declare . . . [their building] a cultural monument,"[70] but based on the prosecutors' own reasoning, Pavlensky also had his lawyer, Olga Dinze, file a complaint that "accused the FSB of illegally replacing the building's doors in 2008." In the complaint, Pavlensky and his lawyer asked the prosecutor general, Yury Chaika, "to hold FSB employees accountable" since "the FSB building" had been "named an object of cultural heritage in 2007." FSB employees had replaced the doors a year later, Pavlensky's lawyer pointed out, and they had not procured the proper "authorization of the Federal Service for the Protection of Cultural Heritage"[71]—an authorization that was required by law. Not only were the FSB employees thus culpable of the very crime with which Pavlensky was charged, but since the doors he burned were not original and had been illegally installed, Pavlensky and his lawyer reasoned, he had not in fact damaged a cultural heritage site at all. On the contrary, through his act of artistic provocation, he had exposed the criminal damage that had been done to Russian culture and heritage by the FSB employees when they changed the doors and, by rhetorical extension, when they had detained and tortured Russia's "outstanding cultural figures" as well—not to mention the damaging and chilling effect that the repressive surveillance regimes of the FSB and its predecessor, the KGB, have had on cultural expression more generally. While all of this absurd legal wrangling transpired, the entrance to the FSB building became its own symbolic sideshow since it was sealed off with corrugated metal, literally placing an "iron curtain"[72] between the public and the KGB's successor, the Russian Security Service.

Amid Pavlensky's disruptive performative mastery of the courtroom, his initial provocative gesture of setting Lubyanka's doors ablaze led to a nomination for the prestigious Russian Innovatsiva Prize, a nomination that Pavlensky accepted and that Mikhail Mindlin, the director of "the state-run National Centre for Contemporary Arts" rejected on the grounds that Pavlensky's piece "involved 'breaches of the law and caused material damage.'"[73] But for a very short period, Pavlensky simultaneously stood as a criminal defendant and as a potential prestigious award recipient all for the same performance piece. Mindlin's rejection of the nomination did not go unanswered. Inspired, one supposes, by Pavlensky's own courage, the jury for the Innovatsiva Prize took offense at Mindlin's ruling, staged a walkout, and ultimately caused the director to cancel the prize altogether rather than risking the possibility that Pavlensky might receive it. While Pavlensky's aesthetically inspired activism may have effectively derailed the Innovatsiva Prize, ultimately his status as a rejected nominee was arguably more valuable than the prize itself.

If Pavlensky's work caused the award system of the Innovatsiva Prize to crash, it was having a similar effect on Russia's legal system, where the performative aspects of his work were still unfolding. The controversy generated by the cancelation of the Innovatsiva Prize increased international interest in Pavlensky's plight in the courtroom, as did the fact he was shuffling back and forth from different courtrooms on different charges: for *Lubyanka's Burning Door* he faced charges of damaging a cultural heritage site, and for his earlier piece *Freedom* he faced charges of vandalism and "petty hooliganism,"[74] after he and other artists built "mock barricades" and burned tires on Saint Petersburg's Maliy Konyushenniy Bridge. These latter actions were conceptualized to demonstrate "support for anti-Russian protesters in the Ukraine"[75] by staging the reenactment of "Kyiv's Maidan protests"[76] that I mentioned in the introduction to this book.

In the public eye, the two trials quickly bled together into one composite performance spectacle where, moving back and forth between courtrooms, Pavlensky adeptly disrupted the Russian legal process. The courtroom proceedings in the latter case where Pavlensky was charged with hooliganism proved to be no less sensational than those of the former. After the prosecution brought in a number of witnesses to testify how offended they were by Pavlensky's pro-Ukrainian performance art on the Maliy Konyushenniy Bridge, Pavlensky paid three prostitutes to join in with the prosecution's witnesses and to testify that they too were offended. Each one of the prostitutes, Pavlensky explained, "was able to rip through the scenery and demonstrate to everyone reality as it is."[77] "They are prostitutes whom I paid so that they would come and testify," he added. "And it is equivalent to the testimony of other witnesses for the prosecution since they have just as much to do with the case. They have exactly the same motive."[78] While the prostitutes added a sensational level of farce to the trial that indicted the entire proceedings, Pavlensky's goals were clearly larger than his own defense: a point that he made abundantly clear when he "rejected the court's proposal to drop the criminal charges against him." He reasoned that he and his lawyer Dmitry Dinze still had witnesses who needed to "be questioned."[79]

The witness Pavlensky most wanted question was a young lawyer by the name of Pavel Yasman. Yasman had been one of the original investigating detectives who had interrogated Pavlensky after he had had been arrested in February 2014 for burning tires on the Maliy Konyushenniy Bridge. But rather than procuring a confession, Yasman found himself on the receiving end of Pavlensky's mesmerizing vision of the role of activist art in society. In fact, "after spending just over four months interacting" with Pavlensky, Yasman considered Pavlensky's vision of political art to be so compelling that he "quit his job at Russia's Investigative Committee and began preparing to become a lawyer."[80] Initially, Yasman had hoped to become part of Pavlensky's own defense team, but since his previous involvement with the case disqualified him, Yasman stood ready instead as a witness for the defense. The fact that Pavlensky proved to be intellectually adroit enough to flip an interrogator was certainly a stunning turn of events, and most accounts of the trial tend to focus on Yasman's flip as Pavlensky's own pièce de résistance, especially since it can be pinpointed to a specific moment in the interrogation when Pavlensky was able to get Yasman to admit that he was merely "a tool" in a power structure that "instumentalizes people and everything" else.[81] But rather than focusing on Yasman's flip, I want to suggest that it was the orchestrated backdrop to his shift in allegiance that was Pavlensky's real accomplishment and that Yasman's flip was merely a by-product (albeit a spectacular one) of a much grander conversion—"a hack," as it were—that took advantage of the vulnerabilities posed by the availability of the most basic surveillance technologies and that redirected them into an affirmation of a radical political aesthetic: one that embraces acts of civil disobedience and disruption and of seeming vandalism and destruction as legitimate forms of activist-artistic expression and as moments of creative resistance to oppressive, authoritarian, or postdemocratic political formations.

One simple fact sets this backdrop into critical relief. Cultural critics can pinpoint the moment of Yasman's shift in allegiance not because they have access to the Investigative Committee's files or Yasman's notes but because Pavlensky secretly recorded the interrogations and then published an edited version of them online as a three-act play.[82] At a conceptual level, the creation of this novel piece of verbatim theater carries slight echoes of McKenzie's "interhacktivity," that, his notion of redirecting the interaction "between humans and computers" in such a way that it "not only takes aim at technical systems, *but also targets social systems*" as well—except in this instance it was the interaction between humans and surveillance technologies that was redirected, and yet redirected with similar effects. Steve Mann has characterized such redirection as "inverse surveillance," or as what he more famously has called "sousveillance." A dialectical concept, Mann's notion of sousveillance, which literally means to "watch from below," is the direct antinomy of surveillance (which means to watch from above), and like McKenzie's sense that interhactivity "not only takes aim at technical systems, *but also targets social systems*," sousveillance targets both surveillance regimes and the sociopolitical systems that cultivate and benefit from them—or at least this is Mann's argument. As a variation of what Mann calls sousveillance, Pavlensky's intervention, his surreptitious recording of his interrogation, targeted both Russia's security system and its legal system: two of the central pillars of the Russian authoritarian state.

In one sense, the term "sousveillance" is a conceptual marker of the vulnerabilities that surveillance regimes inevitably produce, and so it always carries with it a reference to the hierarchies of political power and to the asymmetrical use of surveillance technologies against the powerful by the weak. In Mann's original conception, for example, sousveillance referred to "the recording or monitoring of a high ranking official by a person of lower authority,"[83] and in this regard, there is much overlap between Mann's notion of acts of sousveillance and Pavlensky's notion of his own artistic practice, which he has characterized as "a deliberate attempt to recapture the initiative for the downtrodden 'little man' in the face of the grinding machinery of Russian state power."[84] But where the American media theorist and the Russian actionist part ways is in Mann's belief that sousveillance ultimately is a strategy for democratic accountability, transparency, and objectivity[85]—not to mention his rather questionable belief that an ever-widening distribution of the means of surveillance will counteract the surveillance society's reinforcement of postdemocracy. Perhaps because he works from within a Russian political context, Pavlensky's vision is far less optimistic about the willingness of the powerful to relinquish the tools of authority, and so the aim of his artistic practice, as he amply demonstrated in both of the legal proceedings against him, "is to 'suck the authorities into his art' and deprive them, at least temporarily, of the ability to control events."[86] The goal is to provoke and to lure authorities into an engagement on terms with which they are unfamiliar, or, to be more precise, on the terms of a tactical creative activism, the political aesthetic of which owes allegiance to neither conventional notions of art nor of politics. To borrow Noah Sneider's somewhat clichéd metaphor, Pavlensky "paints with the mechanisms of power,"[87] and inasmuch as his actions abide by neither conventional notions of art nor of politics, they point toward the kind of creative activism that throughout this chapter I have characterized as performance postdemocracy.

While acts of sousveillance may contribute to his goals—and Pavlensky has been careful to disseminate photos and video footage of all of his major interventions, particularly when representatives of authority intervene—they figure only as one among many elements on the genuinely radical palette from which Pavlensky draws for his creative activism. In this respect, amid the sensation of flipping his interrogator and of surreptitiously recording his own interrogation, it is easy to lose sight of something far more important: the moment of creative sedition embraced in the play that Pavlensky produced from the recorded transcripts of his interrogation. In that play, Pavlensky offers an impassioned defense of—as well as an uncompromising assertion of the need for—artistically unconventional and politically radical works that do not shy away from acts that authorities may interpret as vandalism, acts that are the product of a willingness to take on postdemocratic formations by doing "a bit of damage." Indeed, such works defy even conventional understandings of performance, particularly as they pertain to established theatrical practice. "I have never worked with performance," Pavlensky claims: "If you imagine a line, where at one end there is opera (as a means of communication), and at the other there is a terrorist attack (as a means of communication), then in terms of the scriptedness of the gesture, performance will be closer to opera, and actionism to terrorism."[88] That Pavlensky identifies with the aesthetic radicalism of actionism places him within a rich creative tradition of Russian artistic militancy. That he characterizes his artistic practice as a form of terrorism places us back in the courtroom where he had demanded that he be charged with terrorism.

There, when all was said and done, Pavlensky had embarrassed the legal authorities so badly that even with convictions against him in both cases they balked at sentencing him to prison, and so after seven months' detention while facing charges on two separate counts, Pavlensky was released. He emerged from jail only to add a critical coda to his performance, telling reporters that he had seen "the machine from inside" and that he had seen "in concentrated form the surveillance service." He told reporters that he had "clashed with police surveillance" every day, that "prison is a laboratory," and that "you can see the same thing in so-called life at liberty. The same mechanisms, the same methods of control and compulsion that have proven effective in prison," he argued, "are used on the masses."[89] While I do want to note that public statements like these reaffirm the arguments about prisons that I made earlier in this book, Pavlensky's recognition that "prison is a laboratory" for refining coercive methods of surveillance that then can be widely disseminated to control the public adds an important nuance to those earlier discussions. But one additional observation is in order: where there is a laboratory, the outcomes are uncertain, and vulnerabilities exist that might be exploited. Indeed, in such contexts repression and its radical antagonistic antinomy walk together. Both develop new methods. "The process of determining the limits and forms of political art is still going on," Pavlensky argued as he was released. "Even in prison, that did not stop."[90] Though stated in general terms about political art, the immediate point of reference for Pavlensky's comments was his piece *Lubyanka's Burning Door*, which rather than ending with his arrest on November 9 continued throughout his incarceration and trial. Even while he was in jail, it did not stop, and there the piece evolved into one of the most compelling works of antisurveillance political art in last decade.

Assembling the Pieces of Performance Postdemocracy

Perhaps a moment of reflection is in order because we have moved a long way from the Electronic Disturbance Theater. Granted, almost twenty years separate Pavlensky's *Lubyanka's Burning Door* from the Electronic Disturbance Theater's creative intervention in support of the Zapatistas, and this does not even begin to address the vast cultural divide separating Vladimir Putin's Russia in 2015–2016 and Ernesto Zedillo's Mexico in the late 1990s. And although I have suggested that we view the former through the distant conceptual lens of the latter, ultimately I am not seeking equivalencies so much as I am piecing together complementary parts of a larger, emerging political aesthetic. So in juxtaposing these two very different cultural and historical moments of creative activism—or in placing them in the larger context of coordinated discussions of the work of Christoph Faulhaber and of Pyotr Pavlensky—I am looking for something more than conceptual similarities that might unite them under the banner of Badiou's call for a new militancy in theatrical practice or even under the banner of what I have called performance postdemocracy. Rather I seek an equation that will link "electronic civil disobedience" with its contemporary material counterparts and that, in doing so, will yield something new—not in the naive sense of a claim to originality but in the sense of being a combination that can point toward the complex militant forms of creative antisurveillance activism that will be required in the years to come: the forms that will constitute an effective radical antagonism for the twenty-first century and a creative activism that then legitimately might be called theater's new militancy because it recognizes and responds to—

indeed, is inevitably produced by—the new political realities of advanced surveillance societies and of postdemocratic formations.

Like Pavlensky's own aesthetic praxis, such creative activism has little concern with the institutions of art per se. The focus of its disruptiveness is elsewhere. If it is to function dialectically and exploit postdemocracy's inevitable vulnerabilities, then it must surface where it is not expected and do so with little regard for institutionalized artistic legitimacy in either the present or the future. As Susan Buck-Morss has argued in her book *Thinking Past Terror*: "The institutions of cultural power are not threatened by what the artist creates, so long as it is done within the authorized, artworld space."[91] Though her referent is spatial, the issue raised by Buck-Morss is not so much about art produced in and for the traditional spaces of theaters, galleries, and museums. The referent in her allusion to authorized art-world spaces concerns the extent to which artists' "critical and creative powers are kept isolated from social effect"[92] irrespective of whether those powers are applied to artistic expression in the theater, in the streets, or in the digital byways of virtual reality.

Supplanting that isolation with political efficacy is the great challenge of art in an era of expansive surveillance and increasingly fortified postdemocracy, and we could do far worse than to conceptualize the art that rises to that challenge at the far end of the "authorized—unauthorized" binary that informs Buck-Morss's argument. The point is that if "the institutions of cultural power are not threatened" by art, neither are the institutions of political power—those entrenched and fortified forces of postdemocracy. In an era of postdemocracy, art that is authorized is most likely art that is already contained. The need is for art that is interested less in pushing the boundaries of art itself than in the radical possibilities of creative political expression that occupies, that squats on, and that hacks into unauthorized space—art that is motivated by a willingness to burn down the doors that regulate the political order of post-democratic formations, art that is willing to be disruptive and to do a "bit of damage."

But if Buck-Morss's arguments are valid, we might legitimately inquire where one might find the unauthorized spaces of vulnerability that postdemocracy inevitably produces and that might be exploited, for there one is also apt to find the potential spaces of an effective creative activism. Like the hacker's tap of the return key, they are located, conceptually at least, in unauthorized command keystrokes and in the digital underbelly of denial-of-service (DoS) notifications—somewhere in the virtual realms just outside of your server's firewall. Like the activist-arsonist's flick of the lighter, they are located in the calculated control burn of incendiary demonstrations and in the radical return of "pyrotechnics" to its Greek roots, *pyro* meaning "fire" and *tekhnikos* meaning "made by art"[93]—somewhere just outside of the Russian Security Services Building. Like the graffiti artist's vandalizing of private property, so too are they located in interventions that are without license or permit—interventions like the infamous 215-foot penis that the Russian actionist and "underground, oppositionist art group . . . Viona (the Russian word for 'war')" painted on the Liteiny drawbridge in St. Petersburg in 2010 so that when it was raised a giant phallus overshadowed the Saint Petersburg FSB (former KGB) headquarters.[94]

That intervention, of course, echoed a similar one undertaken in 1991 in Moscow when members of ETI, which is a Russian acronym for "Expropriation of the Territory of Art," spontaneously appeared in Red Square, laid down, "and arranged their bodies to spell the word ХУЙ (COCK)."[95] They remained there until they were forcibly removed, a result that symbolically enacted the emasculation and expropriation of artistic practice. Not unlike Christoph Faulhaber's later interventions outside of the U.S. embassies in Berlin and Warsaw, ETI leaders like Anatoly Osmolovsky saw this act as gesture toward recovery of public spaces and claimed, for example, that its goal "was the de-sacralisation of Red Square and its transformation into a people's place."[96] But whatever its goals, so too was it a concession.

Just as the movement's name, Expropriation of the Territory of Art, takes a sarcastic swipe at the appropriation and neutralization of artistic practice by the state,[97] Osmolovsky later admitted that the action was the result of dwindling options for artistic expression: "The only thing left to us," he conceded, "were the streets."[98] Here too we find the faint echo of that pressing question, what other options were available? And as the Russian security forces made abundantly clear, the streets were not really available and were subject to control. Not even the streets were authorized art space. But maybe there is space here for revision, at least a revision of the darker vision of Brecht's "Street Scene" with which I began this book. For in moments of political urgency, the streets—like the information superhighway—might offer unauthorized avenues of vulnerability that can be exploited, and they thus can become powerful sites of resistance. Perhaps a recognition of that possibility ends this book on a more optimistic note about the streets than that with which it began or than that which it explored in the chapter on undercover police operating in activist circles. But if those possibilities do exist, they come with a price.

In 2013, some twenty-two years after ETI's irreverent stunt, Pavlensky entered Red Square himself and, rather than dallying about with words, stripped, sat down, and nailed his scrotum to the paving stones—an act that quickly garnered international attention. Foreshadowing comments he would later make with respect to *Lubyanka's Burning Door*, Pavlensky justified his act public self-mutilation not with the claim that the streets were the only thing left to us but rather that the streets like everywhere else had become "one big prison" and part of a vast surveillance complex where the government steals "from the people" and uses "the money to grow and enrich the police apparatus and other repressive structures."[99] For many in the West, this act was a baffling show of masochism, but for Russians there was something all too familiar in this provocative act of defiant self-mutilation. The Russian cultural critic Marat Guelman, for example, placed it within a long Russian dissident tradition of radical antagonism toward authoritarianism that "developed in Russian prisons, where the inmates nail their scrotums to stools when they have lost all hope of being listened to by the prison authorities."[100] With his provocative and shocking act in Red Square, Pavlensky brought this sensibility out into the open, extending it Russian society more generally. Unlike the playful and canivalesque undercurrent of the interventions by ETI in the 1990s or of Viona just three years earlier, Pavlensky's act signaled the beginning of a much more serious, consequential, and sober brand of actionism.

His was an act of radical antagonism that left playfulness behind and took aim at the stark political realities of a postdemocratic order that had risen from the rubble of the former Soviet Union's bleak authoritarianism before democracy had even had a chance to seed. "The point of this action," Guelman argued, "is to show society and the opposition that we have lost, that the battle is over: they've imprisoned us all and nailed us to the ground."[101] Pavlensky's act completely baffled the security forces at Red Square. Not knowing what to do, "they draped him with a white sheet, turning him into a fleeting reflection of Gandhi"[102] until someone could be located who could dislodge the nail and hence dislodge Pavlensky himself, who was poised to become a permanent fixture in the Russian cultural-political imagination. Here as elsewhere, the blunt materiality of Pavlensky's creative activism—which has been a constant in his work since he sewed his lips shut in 2012 "outside St Petersburg's Kazan Cathedral" in support of Pussy Riot[103]—is clear reminder that radical antagonism, exploitation of power's vulnerabilities, or effective creative resistance to repressive surveillance regimes do not require technological wizardry as their vanguard. But technological wizardry and blunt materiality might walk hand in hand into the unauthorized spaces of resistance that lie before us. In order to do so, they will need a bearing that not only identifies unauthorized spaces but that, like Pavlensky's own bearing, also locates the creative imagination and strength to commandeer them, and they will require brazen tactical aesthetics and acts of uncommon resolve.

Without a doubt, that walk demands as much courage as it does street-smart artistic creativity—maybe more. However much as we may marvel at the bold creativity of activist-artists like Faulhaber or Pavlensky or of collectives like the Critical Art Ensemble or the Electronic Disturbance Theater, however much we may celebrate the eruption of art in unauthorized spaces, their creative activism is the product of a force of conviction—a radical antagonism—that outweighs concern for existing legal statutes. Savvy though they may be with regard to the intricacies of the law or with regard to the machinery of the legal system more generally, the work of artists like Pavlensky, Faulhaber, the Critical Art Ensemble, or the Electronic Disturbance Theater puts them at substantial risk—at risk of arrest and incarceration, at risk of lengthy and personally devastating legal processes, and at risk of physical harm. ("The convoy guards of the Moscow City Court," for example, assaulted Pavlensky.)[104] So while we might celebrate the courage of their imagination, we must not lose sight of the fact that artists like these possess a rare courage with regard to the personal consequences they potentially pay for their interventions. Maybe this is the same courage that makes their art possible.

Maybe too it is the kind of courage that the surveillance society and the era of postdemocracy require for artistic practice that is of consequence. Maybe in this era, it is the kind of courage that separates art from the institutions and commercial industry of art. If this is the case, then performance postdemocracy is performance in an era when the rules of the game have in fact changed. And the new rules? To echo Baudrillard once again, they are indeed fierce. A decade ago Torin Monahan, writing an essay entitled "Countersurveillance as Political Intervention?" focused on the creative activism of early antisurveillance groups like the Institute for Applied Autonomy, ®™ark, and the Surveillance Camera Players—all of whom he lauded but none of whom, he conceded, ever succeeded in moving beyond "*symbolic* conflict" into actual "*physical* confrontation." Settling instead for "acts of symbolic resistance with the intention of raising public awareness about modern surveillance regimes,"[105] they remained unsuccessful, Monahan argued, at "moving critiques of surveillance beyond the level of the individual to their larger institutional and political origins."[106] Here again, in a moment of critical disappointment with what artists were unable to achieve we discover a longing for what art might be and in particular for what countersurveillance art that is motivated by the courage to engage in physical confrontation might actually achieve: a hack, a disturbance, even a bit of damage that discards the quixotic and risks the confrontational.

Looking, in the closing lines of this book, with a critical eye at the interventions by artists like Faulhaber and Pavlensky, I know that some might legitimately ask what these artists actually accomplished and that some might question whether their interventions, however provocative they might be, ever really rise to the occasion and become something more than symbolic gestures. Granted, there is little that an individual artist or a specific ensemble or collective might achieve on his, her, or their own. But the choice is, I think, between symbolic gesture and exemplary act. Theirs is, I think, a choice to walk the walk and lead by example, leaving us not with symbolism so much as with the looming question: who has the courage to follow a similar path? And art should encourage us. What other options remain?

Notes

Introduction

1. John Rodden, *Repainting the Little Red Schoolhouse: A History of East German Education* (Oxford: Oxford University Press, 2002), 419.

2. Amy Schwartz, ". . . And in America," *Washington Post*, December 8, 1992. http://www.washingtonpost.com/archive/opinions/1992/12/08/and-in-america/1daf3a9f-110d-4484-8e09-a4475ba98c03/ (accessed October 12, 2015).

3. Schwartz, "And in America."

4. Friederike Eigler, "At the Margins of East Berlin's 'Counter-culture': Elke Erb's Winkelzüge and Gabriele Kachold's zügel los," *Women in German Yearbook* 9 (1993): 148.

5. Alain Badiou, *Rhapsody for the Theatre*, trans. Bruno Bosteels (New York: Verso, 2013), 117–118.

6. Walter Benjamin, "What Is Epic Theatre?," in *Illuminations*, trans. Harry Zohn, ed. Hannah Arendt (New York: Schocken Books, 1968), 150.

7. See, for example, Massoud Hayoun, "What Happens during a Deportation Raid in the U.S.?," *Aljazeera*, February 15, 2017. http://www.aljazeera.com/indepth/features/2017/02/immigration-deportation-raid-170214213550603.html (accessed March 7, 2017).

8. Walter Benjamin, "The Work of Art in the Age of Mechanical Reproduction," in *Illuminations*, 241.

9. Elise Morrison, *Discipline and Desire: Surveillance Technologies in Performance* (Ann Arbor: University of Michigan Press, 2016), 69.

10. Jonathan Capehart, "Feidin Santana, Hero of North Charleston," *Washington Post*, April 9, 2015. http://www.washingtonpost.com/blogs/post-partisan/wp/2015/04/09/feidin-santana-hero-of-north-charleston/ (accessed October 12, 2015).

11. Capehart, "Feidin Santana."

12. Douglas Starr, "What Your Cell Phone Can't Tell the Police," *New Yorker*, June 26, 2014. http://www.newyorker.com/news/news-desk/what-your-cell-phone-cant-tell-the-police (accessed April 19, 2015).

13. Kai Biermann, "Betrayed by Our Own Data," *Zeitmagazin*, March 10, 2011. http://www.zeit.de/digital/datenschutz/2011-03/data-protection-malte-spitz (accessed October 12, 2015).

14. Morrison, *Discipline and Desire*, 69.

15. Jason Farman, *Mobile Interface Theory* (New York: Routledge, 2012), 6.

16. Marshall McLuhan, *Understanding Media: The Extensions of Man*, 2nd ed. (New York: New American Library, 1964), 158. I am indebted to Philip Auslander's discussion of this passage in the opening chapter of his book *Liveness* (New York: Routledge, 1999).

17. Bertolt Brecht, "The Street Scene," in *Brecht on Theatre*, trans. John Willett (New York: Hill and Wang, 1964), 122.

18. Motorola, "Eyes on the Street: How Wireless Video Solutions Are Transforming Public Safety," Schaumburg, Il: Motorola, 2008. 2.

19. Brecht, "The Street Scene," 121.

20. Farman, *Mobile Interface Theory*, 46.

21. Farman, *Mobile Interface Theory*, 6.

22. Malcolm McCullough, *Ambient Commons* (Cambridge, MA: MIT Press, 2013), 197.

23. McCullough, *Ambient Commons*, 21.

24. McCullough, *Ambient Commons*, 95.

25. McCullough, *Ambient Commons*, 190.

26. Auslander, *Liveness*, 53.

27. Jordan Crandall, "An Actor of the Street: Events, Agencies, and Gatherings," *SubStance* 40.3 (2011): 49.

28. Evgeny Morozov, *To Save Everything, Click Here* (New York: Public Affairs, 2013), 183.

29. Morozov, *To Save Everything*, 183.

30. Beth Blauer, "Why Data Must Drive Decisions in Government," in *Beyond Transparency* (San Francisco: Code for America Press, 2013), 199.

31. Tim O'Reilly, "Open Data and Algorithmic Regulation," in *Beyond Transparency*, 291.

32. O'Reilly, "Open Data," 291.

33. See Evgeny Morozov, "The Rise of Data and the Death of Politics," *Guardian*, July 20, 2014. http://www.theguardian.com/technology/2014/jul/20/rise-of-data-death-of-politics-evgeny-morozov-algorithmic-regulation/print.

34. Morozov, *To Save Everything*, 184.

35. Mark Andrejevic, "Ubiquitous Surveillance," in *Routledge Handbook of Surveillance Studies*, ed. Kristie Ball, Kevin D. Haggerty, and David Lyon (New York: Routledge, 2012), 92.

36. Andrejevic, "Ubiquitous Surveillance," 93.

37. Andrejevic, "Ubiquitous Surveillance," 94.

38. Pete Fussey and Jon Coaffee, "Urban Spaces of Surveillance," in Ball, Haggerty, and Lyon, *Surveillance Studies*, 206.

39. Fussey and Coaffee, "Urban Spaces of Surveillance," 204.

40. McCullough *Ambient Commons*, 243.

41. Fussey and Coaffee, "Urban Spaces of Surveillance," 204.

42. Fussey and Coaffee, "Urban Spaces of Surveillance," 205.

43. Fussey and Coaffee, "Urban Spaces of Surveillance," 205.

44. Crandall, "Actor of the Street," 49.

45. See, for example, the opening of Joel Schechter's review of Benjamin's *Moscow Diaries* in *Theatre Journal* 39.4 (1987): 528.

46. Benjamin, "Work of Art," 221.

47. Benjamin, "Work of Art," 224.

48. Ralph Eue, "Every Picture Is an Empty Picture." http://films2014.dok-leipzig.de/en/film.aspx?ID=7300 (accessed October 15, 2015).

49. Alex Greenberger, "Pyotr Pavlensky Convicted of Vandalism in Moscow for Pro-Ukraine Protest," *Artnews*, May 15, 2016. http://www.artnews.com/2016/05/19/pyotr-pavlensky-convicted-of-vandalism-in-moscow-for-pro-ukraine-protest/ (accessed September 15, 2016).

Chapter 1

Epigraph: Alain Badiou, "A Short Philosophical Treatise," in *Rhapsody for the Theatre*, 55.

1. David Lyon, *Christians and Sociology: To the Challenge of Sociology . . . a Christian Response* (Downers Grove, IL: InterVarsity Press, 1975), 7.

2. Lyon, *Christians and Sociology*, 41.

3. Lyon, *Christians and Sociology*, 8.

4. Lyon, *Christians and Sociology*, 66.

5. Lyon, *Christians and Sociology*, 68.

6. David Lyon, *The Electronic Eye: The Rise of the Surveillance Society* (Minneapolis: University of Minnesota Press, 1994), ix.

7. David Lyon, *Surveillance Studies: An Overview* (Malden, MA: Polity Press, 2007), 3.

8. Eric Stoddart, *Theological Perspectives on a Surveillance Society* (Farnham, UK: Ashgate, 2011), 30.

9. Lyon, *Christians and Sociology*, 11.

10. William Gibson, *Burning Chrome* (New York: Harper Voyager, 2003), 199.

11. Muriel Spark, *The Abbess of Crewe* (New York: New Directions Books, 1974), 26.

12. David Lyon, *Surveillance Society* (Buckingham: Open University Press, 2001), 3.

13. Lewis MacLeod, "Matters of Care and Control: Surveillance, Omniscience, and Narrative Power in *The Abbess Of Crewe* and *Loitering with Intent*," *Modern Fiction Studies* 54.3 (2008): 578.

14. Andrew Sofer, *Dark Matter: Invisibility in Drama, Theatre and Performance* (Ann Arbor: University of Michigan Press, 2013), 105.

15. Sofer, *Dark Matter*, 105.

16. Arthur Miller, cited in Christopher Bigsby, *Arthur Miller: A Critical Study* (New York: Cambridge University Press, 2000), 310.

17. Sofer, *Dark Matter*, 106.

18. I return to Bentham later in this chapter.

19. Christopher Bigsby, afterword to *The Archbishop's Ceiling* by Arthur Miller (London: Methuen, 1984), 92.

20. Bigsby, afterword, 92.

21. Bigsby, afterword, 92–93.

22. Russel P. Moroziuk, "The Role of Atheism in Marxian Philosophy," *Studies in Soviet Thought* 14.3–4 (1974): 204.

23. Richard Comstock, "The Marxist Critique of Religion: A Persisting Ambiguity," *Journal of the American Academy of Religion* 44.2 (1976): 329.

24. Bertrand Russell, *The Value of Free Thought* (New York: Haldeman-Julius Publications, 1944), 21.

25. David Lyon, *Karl Marx: A Christian Assessment of His Life and Thought* (Downers Grove: InterVarsity Press, 1979), 52.

26. Lyon, *Karl Marx*, 69.

27. Miller, *The Archbishop's Ceiling*, 79.

28. Miller, *The Archbishop's Ceiling*, 15.

29. Miller, *The Archbishop's Ceiling*, 88.

30. Miller, *The Archbishop's Ceiling*, 69.

31. Bigsby, *Arthur Miller*, 309.

32. Karl Marx, "A Contribution to the Critique of Hegel's 'Philosophy of Right,'" in *The Young Hegelians: An Anthology*, ed. Lawrence Stepelevich (London: Cambridge University Press, 1983), 310.

33. Stoddart, *Theological Perspectives*, 30.

34. Lyon, *Christians and Sociology*, 10.

35. Astrit Schmidt-Burkhardt, "The All-Seer: God's Eye as Proto-surveillance," in *CTRL [SPACE]: Rhetorics of Surveillance from Bentham to Big Brother*, ed. Thomas Levin, Ursula Frohne, and Peter Weibel (Cambridge, MA: MIT Press, 2002), 29.

36. Lyon, *The Electronic Eye*, 3.

37. Jessica Riskin, "The Divine Optician," *American Historical Review* 116.2 (2011): 360.

38. Schmidt-Burkhardt, "The All-Seer," 20.

39. Schmidt-Burkhardt, "The All-Seer," 23.

40. Schmidt-Burkhardt, "The All-Seer," 31.

41. Farman, *Mobile Interface Theory*, 37.

42. Farman, *Mobile Interface Theory*, 37.

43. MacLeod, "Matters of Care," 578.

44. MacLeod, "Matters of Care," 578.

45. Louis Althusser, "Ideology and Ideological State Apparatuses," in *Lenin and Philosophy and Other Essays*, trans. Ben Brewster (New York: Monthly Review Press, 2001), 89.

46. Althusser, "Ideology," 118.

47. Althusser, "Ideology," 118.

48. Judith Butler, *The Psychic Life of Power: Theories in Subjection* (Stanford, CA: Stanford University Press, 1997), 106.

49. Althusser, "Ideology," 118.

50. David Lyon, *Identifying Citizens: ID Cards as Surveillance* (Cambridge: Polity Press, 2009), 40.

51. Lyon, *Identifying Citizens*, 41.

52. Lyon, *Identify Citizens*, 58.

53. Butler, *Psychic Life of Power*, 107.

54. Lyon, *Surveillance Studies*, 185.

55. Althusser, "Ideology," 118.

56. Butler, *Psychic Life of Power*, 109.

57. Althusser, "Ideology," 120.

58. Butler, *Psychic Life of Power*, 110.

59. Butler, *Psychic Life of Power*, 110.

60. Althusser, "Ideology," 120.

61. Althusser, "Ideology," 121.

62. Butler, *Psychic Life of Power*, 112.

63. Schmidt-Burkhardt, "The All-Seer," 23.

64. Adam Kotsko, "The Failed Divine Performative: Reading Judith Butler's Critique of Theology with Anselm's *On the Fall of the Devil*," *Journal of Religion* 88.2 (2008): 218.

65. Lyon, *Christians and Sociology*, 10.

66. Jacques Ellul, *The Technological Society* (New York: Vintage Books, 1964), 264.

67. David Lyon, "Whither Shall I Flee? Surveillance, Omniscience and Normativity in the Panopticon," *Christian Scholar's Review* 14.3 (1995): 308.

68. David Lyon, "Bentham's Panopticon: From Moral Architecture to Electronic Surveillance," *Queen's Quarterly* 98.3 (1991): 599.

69. Lyon, *The Electronic Eye*, 63.

70. Lyon, "Bentham's Panopticon," 615.

71. Stoddart, *Theological Perspectives*, 42.

72. David Lyon, *Jesus in Disneyland* (Cambridge: Polity Press, 2000), 66.

73. Mary Dery, *Escape Velocity* (New York: Grove Press, 1996), 50.

74. Michel Foucault, "Panopticism," in *The Surveillance Studies Reader*, ed. Sean Hier and Joshua Greenberg (New York: Open University Press, 2007), 70–71.

75. Lyon, "Bentham's Panopticon," 616.

76. Foucault, "Panopticism," 70.

77. Jeremy Bentham, *An Introduction to the Principles of Morals and Legislation* (Buffalo: Prometheus Books, 1988), 193.

78. Miran Božovič, "'An Utterly Dark Spot': The Fiction of God in Bentham's Panopticon," *Qui Parle* 8.2 (1995): 86.

79. Jeremy Bentham, "Panopticon: Preface, Letters and Postscript," in *The Works of Jeremy Bentham*, vol. 4, ed. J. Bowring (Edinburgh: William Tait, 1838–43), 179. This passage and the previously cited passage from Bentham in note 77 were both cited in Božovič, "An Utterly Dark Spot."

80. Marx, "Critique of Hegel's 'Philosophy of Right,'" 310.

81. John Patrick Shanley, *Doubt: A Parable* (New York: Theatre Communications Group, 2005), 12.

82. Shanley, *Doubt*, 13.

83. Shanley, *Doubt*, 15.

84. Elizabeth Cullingford, "Evil, Sin, or Doubt: The Dramas of Clerical Abuse," *Theatre Journal* 62.2 (2010): 262.

85. Cullingford, "Evil, Sin or Doubt," 261.

86. Shanley, *Doubt*, 12–13.

87. Shanley, *Doubt*, 58.

88. Shanley, *Doubt*, 58.

89. Stoddart, *Theological Perspectives*, 7.

90. Stoddart, *Theological Perspectives*, 7.

91. Stoddart, *Theological Perspectives*, 2.

92. Lyon, *The Electronic Eye*, 5.

93. Lyon, *The Electronic Eye*, 13.

94. David Lyon, "Resisting Surveillance," in Hier and Greenberg, *The Surveillance Studies Reader*, 373.

95. Elias Groll, "The Skeletons in Benedict's Closet," February 22, 2013. http://foreignpolicy.com/2013/02/22/the-skeletons-in-benedicts-closet/ (accessed July 1, 2015).

96. Groll, "Skeletons in Benedict's Closet."

97. Jamie Doward, "Pope 'Obstructed' Sex Abuse Inquiry," *Guardian*, April 24, 2005. http://www.theguardian.com/world/2005/apr/24/children.childprotection (accessed July 1, 2015).

98. Schmidt-Burkhardt, "The All-Seer," 18.

99. Hebrews 11.1 (KJV).

Chapter 2

Epigraph: J. Edgar Hoover, cited in Bud Schultz and Ruth Schultz, *The Price of Dissent: Testimonies to Political Repression in America* (Berkeley: University of California Press, 2001), 220.

1. Michael Kirby, "The New Theatre," *Tulane Drama Review* 10.2 (1965): 23.

2. Allan Kaprow, "The Legacy of Jackson Pollock," in *Essays on the Blurring of Art and Life* (Berkeley: University of California Press, 2003), 7–9.

3. Barbara Haskell, *Blam! The Explosion of Pop, Minimalism and Performance, 1958–1964* (New York: W. W. Norton, 1984), 17.

4. Haskell, *Blam!*, 17.

5. Haskell, *Blam!*, 17.

6. Emily Colborn-Roxworthy, "Role-Play Training at a 'Violent Disneyland': The FBI Academy's Performance Paradigms," *TDR* 48.4 (2004): 91. italics mine).

7. David Cunningham, *There's Something Happening Here: The New Left, The Klan and FBI Counterintelligence* (Berkeley: University of California Press, 2004), 8.

8. Michael Kirby, "Happenings: An Introduction," in *Happenings* (New York: Dutton, 1965), 17.

9. François Vidocq, *Memoirs of Vidocq: Principal Agent of the French Police* (Philadelphia: Carey, Hart & Co., 1834), 6–9.

10. Gary Marx, *Undercover: Police Surveillance in America* (Berkeley: University of California Press, 1988), 18.

11. Marx, *Undercover*, 18.

12. Richard Schechner, *Performance Studies: An Introduction*, 3rd ed. (London: Routledge, 2013), 119.

13. Loch K. Johnson, "Congressional Supervision of America's Secret Agencies: The Experience and Legacy of the Church Committee," *Public Administration Review* 64.1 (2004): 5.

14. Johnson, "Congressional Supervision," 6.

15. Johnson, "Congressional Supervision," 6.

16. Cunningham, *There's Something Happening Here*, 6.

17. Pnina Lahav, "Theater in the Courtroom: The Chicago Conspiracy Trial," *Law and Literature* 16.3 (2004): 384.

18. Lahav, "Theatre in the Courtroom," 408–409.

19. David J. Garrow, "FBI Political Harassment and FBI Historiography: Analyzing Informants and Measuring the Effects," *Public Historian* 10.4 (1988): 9.

20. Cunningham, *There's Something Happening Here*, xiii.

21. Cunningham, *There's Something Happening Here*, xiii.

22. Jean-Jacques Lebel, "Notes on Political Street Theatre, Paris: 1968, 1969," *TDR* 13.4 (1969): 112.

23. John McMillian, "Garbage Guerilla." http://www.webcitation.org/5zvq45u2Z (accessed November 15, 2013).

24. See, for example, the anonymously published history "Black Mask & Up Against the Wall Motherfucker" (no publisher, no date).

25. Andrew Hunt, "'When Did the Sixties Happen?': Searching for New Directions," *Journal of Social History* 33.1 (1999): 150.

26. Cunningham, *There's Something Happening Here*, xiii.

27. James Ledbetter, "The Day the NYSE went Yippie." http://money.cnn.com/2007/07/17/news/funny/abbie_hoffman/index.htm (accessed November 12, 2013).

28. Vincent Cannato, *The Ungovernable City* (New York: Basic Books, 2002), 221.

29. Cannato, *The Ungovernable City*, 222.

30. Garrow, "FBI Political Harassment," 6.

31. David Cunningham, "Understanding State Responses to Left- versus Right-Wing Threats: The FBI's Repression of the New Left and the Ku Klux Klan," *Social Science History* 27.3 (2003): 329–330.

32. See Gary May, *The Informant: The FBI, the Ku Klux Klan and the Murder of Viola Liuzzo* (New Haven: Yale University Press, 2005).

33. Cunningham, "Understanding State Responses," 330.

34. Cunningham, "Understanding State Responses," 330.

35. Kenneth O'Reilly, *"Race Matters": The FBI's Secret File on Black America, 1960–1972* (New York: Free Press, 1989), 198. Also cited in Cunningham, "Understanding State Responses," 331.

36. Cunningham, "Understanding State Responses," 330–331.

37. Marx, *Undercover*, 99.

38. PEN America, *Chilling Effect: NSA Surveillance Drives U.S. Writers to Self-Censor* (New York: PEN America Center, 2013), 4. (I want to thank Kate Bredeson for bringing this report to my attention.)

39. Marx, *Undercover*, 130.

40. Diana Welch, "The Informant/Revolutionary to Rat: The Uneasy Journey of Brandon Darby," *Austin Chronicle*, January 23, 2009. http://www.austinchronicle.com/news/2009-01-23/729400/ (accessed December 4, 2013).

41. Josh Harkinson, "How a Radical Leftist Became the FBI's BBF," *Mother Jones*, September–October 2011. http://www.motherjones.com/politics/2011/08/brandon-darby-anarchist-fbi-terrorism?page=1 (accessed December 5, 2013).

42. Cited in Welch, "The Informant."

43. Cited in Welch, "The Informant."

44. Lisa Fithian, "FBI Informant Brandon Darby: Sexism, Egos, and Lies," *Rag Blog*, March 22, 2010. http://theragblog.blogspot.com/2010/03/lisa-fithian-fbi-informant-brandon.html (accessed December 6, 2013).

45. Harkinson, "Radical Leftist."

46. Welch, "The Informant."

47. Scott Crow, *Black Flags and Windmills and the Common Ground Collective* (Oakland, CA: PM Press, 2011), 167.

48. Crow, *Black Flags and Windmills*, 167.

49. Harkinson, "Radical Leftist."

50. Fithian, "FBI Informant Brandon Darby."

51. David Hanners, "Feds Seek to Shield Informant in Retrial of RNC

Molotov Cocktail Case," *St. Paul Pioneer Press*, March 5, 2009. http://rnc08report.org/archive/832.shtml (accessed December 7, 2013).

52. Harkinson, "Radical Leftist.

53. "Interview with Brandon Darby," *Informant*, dir. Jamie Meltzer, Music Box Films, 2013.

54. Cited in Pamela Colloff, "Entrapped: How a High-Profile Member of Austin's Radical Progressive Community Became an FBI Informant," *Texas Monthly*, February 2009. http://www.texasmonthly.com/story/entrapped (accessed December 6, 2013).

55. Cited in Diana Welch, "Activists Raise Concerns about FBI Informant," *Austin Chronicle*, January 9, 2009. http://www.austinchronicle.com/news/2009-01-09/724178/ (accessed December 7, 2013).

56. Cited in Jake Kring-Schreifels, "'The Informant'—Who Is Brandon Darby," *Indiewood/Hollywoodn't*, September 13, 2013. http://indienyc.com/the-informant-who-is-brandon-darby/ (accessed December 6, 2013).

57. "Interview with Brandon Darby."

58. Harkinson, "Radical Leftist."

59. Austin Informant Working Group, "Brandon Darby: FBI Informant Is Provocateur, Not Hero," *Rag Blog*, January 7, 2009. http://theragblog.blogspot.com/2009/01/brandon-darby-fbi-informant-is.html (accessed December 8, 2013).

60. "Interview with Brandon Darby."

61. Cunningham, *There's Something Happening Here*, 6.

62. Cited in Austin Informant Working Group, "Brandon Darby."

63. John Soma and Stephen D. Rynerson, *Privacy Law* (St. Paul, MN: Thomson/West, 2008), 8.

64. Ruth Gavison, "Privacy and the Limits of Law," *Yale Law Journal* 89.3 (1980): 459.

65. Erving Goffman, preface to *The Presentation of Self in Everyday Life* (Edinburgh: University of Edinburgh Social Sciences Research Centre, 1956).

66. Gary Marx, "Under-the-Covers Undercover Investigations: Some Reflections on the State's Use of Sex and Deception in Law Enforcement," *Criminal Justice Ethics* 11 (1992): 18.

67. Marx, *Undercover*, 148.

68. Marx, *Undercover*, 148.

69. Cited in Marx, *Undercover*, 99.

70. Peter Arnett, "Major Describes Move," *New York Times*, February 8, 1968.

71. "Police Spies Out of Lives: Support Group for Women's Legal Action against Undercover Policing." http://policespiesoutoflives.org.uk/home/ (accessed December 15, 2013).

72. Morrison, *Discipline and Desire*, 83.

73. Rob Evans and Paul Lewis, "Political Activists Sue Met over Relationships with Police Spies," *Guardian*, November 21, 2012. http://www.

theguardian.com/uk/2012/nov/21/met-police-spies-women-undercover (accessed December 18, 2013).

74. Evans and Lewis, "Political Activists Sue Met."

75. Soma and Rynerson, *Privacy Law*, 25.

76. Goffman, *Presentation of Self*, 38–39.

77. Goffman, *Presentation of Self*, preface.

78. Home Affairs Committee, "Minutes of Evidence HC 837," February 5, 2013. http://www.publications.parliament.uk/pa/cm201213/cmselect/cmhaff/837/130205i.htm (accessed December 16, 2013).

79. Home Affairs Committee, "Minutes of Evidence."

80. JUSTICE, *Freedom from Suspicion: Surveillance Reform for a Digital Age* (London: JUSTICE, 2011), 135.

81. Rob Evans and Paul Lewis, "Metropolitan Police Want Secret Court to Hear Police Spy Cases," *Guardian*, October 2, 2012. http://www.theguardian.com/uk/2012/oct/02/metropolitan-police-spy-cases-tribunal?newsfeed=true&guni=Article:in%20body%20link (accessed December 19, 2013).

82. JUSTICE, *Freedom from Suspicion*, 135.

83. http://www.judiciary.gov.uk/Resources/JCO/Documents/Judgments/akj-cpm-judgment-17012013.pdf (accessed December 20, 2013).

84. Mordecai Richler, "James Bond Unmasked," in *Mass Culture Revisited*, ed. Bernard Rosenberg (New York: Van Nostrand Reinhold, 1971), 344.

85. Home Affairs Committee, "Minutes of Evidence."

86. Soma and Rynerson, *Privacy Law*, 46.

87. Eveline Lubbers, *Secret Manoeuvers in the Dark* (London: Pluto Press, 2012), 18.

88. See, for example, their anthology, Kristie Ball and Laureen Snider, eds., *The Surveillance-Industrial Complex: A Political Economy of Surveillance* (New York: Routledge, 2013).

89. See Time Shorrock, *Spies for Hire* (New York: Simon and Schuster, 2008) 9–37, 38–71.

90. "Military-Industrial Complex Speech, Dwight D. Eisenhower, 1961." http://coursesa.matrix.msu.edu/~hst306/documents/indust.html (accessed December 27, 2013).

91. Simon Hattenstone, "Mark Kennedy: Confessions of an Undercover Cop," *Guardian*, March 25, 2011. http://www.theguardian.com/environment/2011/mar/26/mark-kennedy-undercover-cop-environmental-activist (accessed April 24, 2017).

92. See Rob Evans and Paul Lewis, *Undercover: The True Story of Britain's Secret Police* (London: Faber and Faber, 2013), 9–17.

93. Evans and Lewis, *Undercover*, 325.

94. Evans and Lewis, *Undercover*, 295.

95. Ian Birrell, "Philandering Policemen, Secret Courts, and This Insidious Threat to Our Faith in Justice," *Mail On-line*, November 19, 2012. http://

www.dailymail.co.uk/debate/article-2235503/Mark-Kennedy-Philandering-policemen-secret-courts-insidious-threat-faith-justice.html (accessed January 11, 2014).

96. See, for example, "New Film Tells Eco-Protest and Undercover Police Officer Mark Kennedy's Story," *Nottingham Post*, October 4, 2013. http://www.nottinghampost.com/New-film-tells-eco-protest-undercover-police/story-19886542-detail/story.html (accessed January 11, 2014).

97. Evans and Lewis, *Undercover*, 295.

98. Lubbers, *Secret Manoeuvers*, ix.

99. Rob Evans, "Mark Kennedy Advises Firms on Activists," *Guardian*, June 22, 2012. http://www.theguardian.com/uk/undercover-with-paul-lewis-and-rob-evans/2012/jun/22/mark-kennedy-undercover-policing (accessed April 24, 2017).

100. Evans and Lewis, *Undercover*, 294.

Chapter 3

Epigraph: Jeremy Bentham, "Constitutional Code, Book II, Ch. XII, sect. XIV," in *The Works of Jeremy Bentham*, vol. 9 (Edinburgh: Tait, 1843), 493.

1. Joshua L. Dratel, "The Legal Narrative," in *The Torture Papers: The Road to Abu Ghraib*, ed. Karen J. Greenberg and Joshua L. Dratel (Oxford: Oxford University Press, 2005), xxi.

2. Although "Camp X-Ray" was a temporary facility that was replaced by Camp Delta, I use the term throughout this chapter to suggest that the shift to "Camp Delta" was as much about the PR disaster of a name like "Camp X-Ray" as it was about housing prisoners in more permanent facilities.

3. Jacques-Alain Miller, "Jeremy Bentham's Panoptic Device" (1975), trans. Richard Miller, *October* 41 (1987): 4.

4. Michel Foucault, *Discipline and Punish*, trans. Alan Sheridan (New York: Vintage, 1995), 200; italics mine.

5. Miller, "Jeremy Bentham's Panoptic Device," 9.

6. Kevin Haggerty and Richard Ericson, "The Surveillant Assemblage," in Hier and Greenberg, *The Surveillance Studies Reader*, 105.

7. Zygmunt Bauman and David Lyon, *Liquid Surveillance* (Cambridge: Polity Press, 2013), 61.

8. Foucault, *Discipline and Punish*, 201.

9. Colin Crouch, *Post-democracy* (Cambridge: Polity Press, 2004), 19.

10. Crouch, *Post-democracy*, back cover.

11. Crouch, *Post-democracy*, 19.

12. Force-feeding of prisoners on hunger strike is but one example of such injustices. This practice is in violation of international law. See http://talkingpointsmemo.com/news/judge-guantanamo-force-feeding-appears-to-be-illegal-but-i-cant-stop-it.php (accessed July 19, 2013).

13. https://www.aclu.org/infographic/guantanamo-numbers (accessed February 20, 2017).

14. Karen Greenberg, "When Gitmo Was (Relatively) Good," *Washington Post*, January 25, 2009. www.washingtonpost.com/wp-dyn/content/article/2009/01/23/AR2009012302313_pf.html (accessed January 27, 2013).

15. Bentham, "Constitutional Code," 493.

16. Miller, "Jeremy Bentham's Panoptic Device," 9.

17. Foucault, *Discipline and Punish*, 207.

18. Lyon, *Surveillance Studies*, 182.

19. Ryan Claycomb, "*Guantanamo: Honor Bound to Defend Freedom* (Review)," *Theatre Journal* 58.4 (2006): 703.

20. Stephen Reicher and S. Alexander Haslam, "Rethinking the Psychology of Tyranny: The BBC Prison Study," *British Journal of Social Psychology* 45 (2006): 7.

21. Neither the Zmijewski film nor the BBC Prison Experiment was the first to call attention to the Stanford Prison Experiment once again. Both works emerged shortly after the German director Oliver Hirschbiegel finished his 2001 film *Das Experiment* (which was subsequently remade in English by Paul Sheuring as *The Experiment* in 2010). That film differed from the two subsequent works in that rather than re-creating the conditions of the original experiment, it simply offered a fictional narrative roughly based upon the actual events in the Stanford experiment itself.

22. Anthony Downey, "The Lives of Others: Artur Zmijewski's *Repetition*, the Stanford Prison Experiment, and the Ethics of Surveillance," in *Conspiracy Dwellings: Surveillance in Contemporary Art*, ed. Outi Remes and Pam Skelton (Newcastle upon Tyne: Cambridge Scholars Publishing, 2010), 71–72.

23. Downey, "The Lives of Others," 73–74.

24. Reicher and Haslam, "Rethinking the Psychology," 4.

25. Downey, "The Lives of Others," 69.

26. Philip G. Zimbardo, *The Lucifer Effect* (New York: Random House, 2007), 329.

27. Downey, "The Lives of Others," 70.

28. The number seven, rather than Pirandello's *Six Characters*, refers to those who were prosecuted.

29. Zimbardo, *The Lucifer Effect*, 381.

30. Didi Bigo, "Globalized (in)Security: The Field and the Ban-opticon," in *Terror, Insecurity and Liberty: Illiberal Practices of Liberal Regimes after 9/11* (London: Routledge, 2008), 32.

31. Bigo, "Globalized (in)Security," 33.

32. James Schlesinger, *Final Report of the Independent Panel to Review DOD Detention Operations*, August 2004, appendix G, p. 1.

33. Bigo, "Globalized (in)Security," 33.

34. Reicher and Haslam, "Rethinking the Psychology," 8.

35. Reicher and Haslam, "Rethinking the Psychology," 25.

36. Reicher and Haslam, "Rethinking the Psychology," 4.

37. Reicher and Haslam, "Rethinking the Psychology," 30.

38. Philip G. Zimbardo, "On Rethinking the Psychology of Tyranny: The BBC Prison Study," *British Journal of Social Psychology* 45 (2006): 53.

39. Downey, "The Lives of Others," 68.

40. Zimbardo, *The Lucifer Effect*, 352.

41. Caoimhe McAvinchey, *Theatre and Prison* (London: Palgrave Macmillan, 2011), 29–30.

42. Reicher and Haslam, "Rethinking the Psychology," 26.

43. Reicher and Haslam, "Rethinking the Psychology," 26.

44. Simone Brown, *Dark Matters: On the Surveillance of Blackness* (Durham, NC: Duke University Press, 2015), 15.

45. Reicher and Haslam, "Rethinking the Psychology," 6.

46. Zimbardo, *The Lucifer Effect*, 252.

47. Zimbardo, "Rethinking the Psychology," 52.

48. http://www.history.com/this-day-in-history/eisenhower-warns-of-military-industrial-complex (accessed August 28, 2016).

49. Tony Perucci, "Performance Complexes: Abu Ghraib and the Culture of Neoliberalism," in *Violence Performance: Local Roots and Global Routes of Conflict*, ed. Patrick Anderson and Jisha Menon (New York: Palgrave Macmillan, 2009), 358.

50. Jon McKenzie, "Abu Ghraib and the Spectacle of the Scaffold," in Anderson and Menon, *Violence Performance*, 339.

51. http://www.humanrightsfirst.org/our-work/law-and-security/we-can-end-torture-now/the-issue/ (accessed August 31, 2013).

52. MG George R. Fay, *AR 15-6 Investigation of the Abu Ghraib Detention Facility and 205th Military Intelligence Brigade*, 2004, 19.

53. Griff Witte and Renae Merle, "Contractors Are Cited in Abuses at Guantanamo," *Washington Post*, January 4, 2007.

54. Witte and Merle, "Contractors Are Cited."

55. Naomi Wolf, "America's Zombie Prison," March 30, 2013. www.project-syndicate.org/print/how-guantanamo-bay-is-worse-than-alcatraz (accessed August 31, 2013).

56. Wolf, "America's Zombie Prison."

57. Athol Fugard, John Kani, and Winston Ntshona, *The Island* (New York: Theatre Communications Group, 1986), 63.

58. Fugard, Kani, and Ntshona, *The Island*, 55.

59. McAvinchey, *Theatre and Prison*, 3.

60. "Outside the Wire." http://www.outsidethewirellc.com (accessed August 22, 2016).

61. Justin Ailes, "Ancient Plays Provide Forum for Tough Topics," *America's Navy*, June 23, 2011. http://www.navy.mil/submit/display.asp?story_id=61206 (accessed January 27, 2013).

62. Ailes, "Ancient Plays Provide Forum."

63. Bryan Doerries, *The Theater of War: What Ancient Tragedies Can Teach Us Today* (New York: Vintage, 2016), 158.

64. Doerries, *The Theater of War*, 156.

65. Doerries, *The Theater of War*, 197.

66. https://ccrjustice.org/home/get-involved/tools-resources/fact-sheets-and-faqs/solitary-confinement-guantanamo-bay#4 (accessed September 1, 2016).

67. Errol Durbach, "Sophocles in South Africa: Athol Fugard's 'The Island,'" *Comparative Drama* 18.3 (1984): 259.

68. Lyon, *Identifying Citizens*, 4.

69. Lyon, *Identifying Citizens*, 45–46.

70. Lyon, *Identifying Citizens*, 25.

71. Michelle Alexander, *The New Jim Crow: Mass Incarceration in the Age of Colorblindness* (New York: New Press, 2012), 6.

72. Vicky Pelaez, "The Prison Industry in the United States: Big Business or a New Form of Slavery?," *Global Research*, March 31, 2014. http://www.globalresearch.ca/the-prison-industry-in-the-united-states-big-business-or-a-new-form-of-slavery/8289 (accessed August 17, 2016).

73. Peter Wagner and Bernadette Rabuy, "Mass Incarceration: The Whole Pie 2016," *Prison Policy Initiative*, March 14, 2016. http://www.prisonpolicy.org/reports/pie2016.html (accessed August 30, 2016).

74. Adam Gopnik, "The Caging of America: Why Do We Lock Up So Many People?," *New Yorker*, January 30, 2012. http://www.newyorker.com/magazine/2012/01/30/the-caging-of-america (accessed August 19, 2016).

75. Alexander, *The New Jim Crow*, 94–95.

76. http://www.sentencingproject.org/wp-content/uploads/2015/11/Americans-with-Criminal-Records-Poverty-and-Opportunity-Profile.pdf (accessed September 2, 2016).

77. Lauren E. Glaze and Danielle Kaeble, "Correctional Populations in the United States, 2013," *U.S. Department of Justice Bulletin*, December 2014 (NCJ 248479) http://www.bjs.gov/content/pub/pdf/cpus13.pdf (accessed September 2, 2016).

78. Alexander, *The New Jim Crow*, 2.

79. Alexander, *The New Jim Crow*, 2.

80. Brown, *Dark Matters*, 10.

81. Brown, *Dark Matters*, 11.

82. Brown, *Dark Matters*, 31.

83. Letter cited in Brown, *Dark Matters*, 31.

84. Brown, *Dark Matters*, 32.

85. Brown, *Dark Matters*, 42.

86. Duncan Campbell and Richard Norton-Taylor, "Prison Ships, Torture Claims, and Missing Detainees," *Guardian*, June 1, 2008. https://www.the

guardian.com/world/2008/jun/02/terrorism.terrorism (accessed September 4, 2016).

87. Duncan Campbell and Richard Norton-Taylor, "US Accused of Holding Terror Suspects on Prison Ships," *Guardian*, June 1, 2008. https://www.theguardian.com/world/2008/jun/02/usa.humanrights (accessed September 4, 2016).

88. Janelle Reinelt, "Toward a Poetics of Theatre and Public Events: The Case of Stephen Lawrence," *TDR* 50.3 (2006): 73.

89. Reinelt, "Poetics of Theatre," 78.

90. Foucault, *Discipline and Punish*, 200.

91. Brown, *Dark Matters*, 32.

92. Lyon, "Bentham's Panopticon," 600.

93. Lyon, "Bentham's Panopticon," 603.

94. Earl Smith and Angela Hattery, "If We Build It, They Will Come: Human Rights Violations and the Prison Industrial Complex," *Societies without Borders* 2 (2007): 283.

95. Smith and Hattery, "If We Build It," 283.

96. Mechthild Nagel, "Beyond the Pale: Reflections on the Vulnerability of Black Life in the United States," in *Diversity, Social Justice and Inclusive Excellence: Transdisciplinary and Global Perspectives*, ed. Seth N. Asumah and Mechthild Nagel (Albany: SUNY Press, 2014), 73.

97. Smith and Hattery, "If We Build It," 281.

98. Cody Mason, *Too Good to Be True: Private Prisons in America* (Washington, DC: Sentencing Project, 2012), 13.

99. "CCA Made Revenues of $1.79bn in 2015, up from $1.65bn in 2014. Geo Group Made Revenues of $1.84bn, a 9% Increase on the Previous Year." Rupert Neate, "Welcome to Jail Inc: How Private Companies Make Money Off US Prisons," *Guardian*, June 16, 2016. https://www.theguardian.com/us-news/2016/jun/16/us-prisons-jail-private-healthcare-companies-profit (accessed August 17, 2016).

100. Michael Cohen, "How For-Profit Prisons Have Become the Biggest Lobby No One Is Talking About," *Washington Post*, April 28, 2015. https://www.washingtonpost.com/posteverything/wp/2015/04/28/how-for-profit-prisons-have-become-the-biggest-lobby-no-one-is-talking-about/?utm_term=.27c936e0a17b (accessed August 16, 2016).

101. Kendall Bensten, "Money, Not Morals, Drives Marijuana Prohibition Movement," *Open Secrets Blog*, August 5, 2014. https://www.opensecrets.org/news/2014/08/money-not-morals-drives-marijuana-prohibition-movement/ (accessed August 16, 2016).

102. Gopnik, "The Caging of America."

103. Perucci, "Performance Complexes," 364; italics mine.

104. Christopher Petrella, "Private Prisons Currently Exempt from Freedom of Information Act," *Prison Divestment Campaign*, September 25, 2012.

https://prisondivest.com/2012/09/25/private-prisons-currently-exempt-from-freedom-of-information-act/ (accessed August 17, 2016).

105. Alex Friedmann et al., "The Dirty Thirty: Nothing to Celebrate about 30 Years of Corrections Corporation of America," *Grassroots Leadership*, no date. http://grassrootsleadership.org/cca-dirty-30#18 (accessed August 17, 2016).

106. David Dayen, "The True Cost: Why the Private Prison Industry Is about So Much More Than Prisons," *Talking Points Memo*, 2016 (no date). http://talkingpointsmemo.com/features/privatization/two/ (accessed August 17, 2016).

Chapter 4

1. http://www.notbored.org/being-watched.html (accessed June 26, 2012).

2. Brown was arguably the key figure in the Surveillance Camera Players while they were active, and he is certainly the key figure in keeping their legacy alive. An immensely energetic and creative political activist, Brown is a former Fulbright scholar, has a Ph.D. in American literature, and "taught at the Rhode Island School of Design" before cofounding the Surveillance Camera Players with a group of like-minded political activists. Nato Thompson et al., *The Interventionists: User's Manual for the Creative Disruption of Everyday Life* (North Adams, MA: Mass MoCA Publications, 2004), 83.

3. Leah Lievrouw, *Alternative and Activist New Media* (New York: Polity Press, 2011), 78.

4. The original *Newsday* article by Samuel Bruchey was reprinted (and archived) by *The Spokesman-Review* on January 24, 2004. https://news.google.com/newspapers?nid=1314&dat=20040124&id=wsRYAAAAIBAJ&sjid=y_IDAAAAIBAJ&pg=2838,1529790&hl=en (accessed February 22, 2017).

5. http://www.notbored.org/scp-antecedents.html (accessed July 13, 2012).

6. Michael Carter, "Guerrilla Programming of Video Surveillance Equipment," in *We Know You Are Watching* (New York: Factory School, 2006), 21–22. Originally distributed "in Nova Scotia as a hand-bill," Carter's short manifesto is cited by the Surveillance Camera Players as "one of the very first documents of its kind," and it appeared in "the same year that Simon Davies [the founder] of Privacy International wrote a classic anti-surveillance article, the publication of which in *Wired Magazine* constituted, in Davies's own words, "the first indication that CCTV was being directly challenged on privacy grounds."

7. Carter, "Guerrilla Programming," 21.

8. Surveillance Camera Players, "The Surveillance Camera Players . . . ," in *We Know You Are Watching*, 22.

9. Raymond Wacks, *Privacy: A Very Short Introduction* (Oxford: Oxford University Press, 2010), 2.

10. http://www.notbored.org/scp-founding.html (accessed June 15, 2012).

11. "On Antonin Artaud and the Theater of Cruelty," in *We Know You Are Watching*, 185.

12. John McGrath, *Loving Big Brother: Performance, Privacy and Surveillance Space* (New York: Routledge, 2004), 196.

13. Morrison, *Discipline and Desire*, 72.

14. Bill Brown, email correspondence with author, July 14, 2012.

15. Brown, email.

16. Brown, email.

17. "Time in the Shadows of Anonymity: Against Surveillance, Transparency and Globalized Capitalism," in *We Know You Are Watching*, 180.

18. "Time in the Shadows."

19. Gabriella Giannachi, *The Politics of New Media Theatre* (New York: Routledge, 2007), 47.

20. Morrison, *Discipline and Desire*, 77.

21. "God's Eyes on Earth," in *We Know You Are Watching*, 26.

22. http://www.videosurveillance.com/churches.asp (accessed July 19, 2012).

23. Mark 11:15–17.

24. C. William R. Webster, Eric Töpfer, Francisco Klauser, and Charles Raab, "Revisiting the Surveillance Camera Revolution: Issues of Governance and Public Policy," *Information Polity* 16.4 (2011): 298.

25. "Time in the Shadows," 173.

26. Claire Valier, *Crime and Punishment in Contemporary Culture* (New York: Routledge, 2004), 152.

27. Clive Norris and Gary Armstrong, *The Maximum Surveillance Society* (Oxford: Berg, 1999), 91.

28. Norris and Armstrong, *The Maximum Surveillance Society*, 92.

29. Norris and Armstrong, *The Maximum Surveillance Society*, 92.

30. Norris and Armstrong, *The Maximum Surveillance Society*, 92–93.

31. Bill Brown, email correspondence with author, July 17, 2012.

32. *We Know You Are Watching*, 229.

33. See, for example: William Webster, "CCTV Policy in the UK: Reconsidering the Evidence Base," *Surveillance and Society* 6.1 (2009) 10–22.

34. Webster, "CCTV Policy," 10.

35. Webster, "CCTV Policy," 13.

36. Torin Monahan, *Surveillance in the Time of Insecurity* (New Brunswick, NJ: Rutgers University Press, 2010), 138.

37. It is the one performance cited, for example, by Emma Govan, Helen Nicholson, and Katie Normington in their book *Making a Performance* (New York: Routledge, 2007), 128–129.

38. Heather Cameron, "CCTV and (In)dividuation," *Surveillance and Society* 2.2–3 (2004): 137.

39. Cameron, "CCTV and (In)dividuation," 137.

40. "IQ Facial Recognition." http://www.iomniscient.com/index.php?view=article&catid=35%3Apr . . . &tmpl=component&print=1&layout=default&page=&option=com_content (accessed June 16, 2012).

41. Cameron, "CCTV and (In)dividuation," 141.

42. Cameron, "CCTV and (In)dividuation," 141.

43. Luis A. Fernandez and Laura Huey, "Is Resistance Futile? Thoughts on Resisting Surveillance," *Surveillance and Society* 6.3 (2009): 199–200.

44. Lucas D. Introna and David Wood, "Picturing Algorithmic Surveillance: The Politics of Facial Recognition Systems," *Surveillance and Society* 2.2–3 (2004): 183.

45. Margit Sutrop and Katrin Laas-Mikko, "From Identity Verification to Behavior Prediction: Ethical Implications of Second Generation Biometrics," *Review of Policy Research* 29.1 (2012): 22.

46. Sutrop and Laas-Mikko, "Identity Verification," 22.

47. Sutrop and Laas-Mikko, "Identity Verification," 22–23. The issue here is not merely that of a normalization of clandestine gathering of biometric features. As Sutrop and Laas-Mikko observe, "in security contexts" not only is "no consent . . . requested," but frequently people are neither "informed about the intended purpose of the collection of the biometrical data," nor informed who is actually doing the collecting. Nor do they know who ultimately will be given access and under what conditions they will be given access to the data collected and to the predictions based upon that data (23). The increasing prevalence of all-pervasive surveillance is thus facilitated ironically by the diminishing possibility of democratic oversight, transparency, and consent.

48. See, for example, Jon C. Leachtenauer and Ronald G. Driggers, *Surveillance and Reconnaissance Imaging Systems: Modeling and Performance Prediction* (Norwood, MA: Artech House, 2001).

49. Sutrop and Laas-Mikko, "Identity Verification," 27.

50. ADABTS, *Automatic Detection of Abnormal Behavior and Threats in Crowded Spaces (ADABTS)*, "ADABTS WP2 User Needs. D.2.1" (2011), 2. https://www.informationsystems.foi.se/main.php/ADABTS_D2%201_User_Needs_Final.pdf?fileitem=10010654 (accessed May 28, 2012).

51. *ADABTS*, "ADABTS WP2 User Needs," 14.

52. *ADABTS*, "ADABTS WP2 User Needs," 16.

53. *ADABTS*, "ADABTS WP2 User Needs," 16.

54. *ADABTS*, WP2 User Needs," 119.

55. *ADABTS*, WP2 User Needs," 119.

56. See, for example, Lyon, *Surveillance Studies*, 99–108.

57. Detec AS describes itself as a market-leader in surveillance equipment within its home market, Norway. It markets both software and hard-

ware, and is involved in the development of digital video-detection products. http://www.detec.no) (accessed April 24, 2017).

58. TNO is a large security company that is involved in the development of advanced surveillance systems. As it describes itself: "TNO offers outstanding performance and inspiring solutions based on our multidisciplinary and experimental research programmes." (http://www.tno.nl/content.cfm?context=thema&content=innovatiegebied&laag1=893&laag2=910&item_id=910 (accessed February 10, 2013).

59. FOI is the Swedish Defence Research Agency. Its main clients are the armed forces and the Swedish Defence Material Administration. But it also caters to large public and private firms. http://www.foi.se (accessed February 10, 2013).

60. BAE SYSTEMS advertises itself as a "global defense, aerospace and security company." http://globalbrochure.baesystems.investis.com/#/2/ (accessed February 10, 2013).

61. http://www.baesystems.com/our-company-rzz/about-us/our-culture (accessed February 10, 2013).

62. See Jay Stanley, *The Surveillance-Industrial Complex: How the American Government Is Conscripting Businesses and Individuals in the Construction of a Surveillance Society* (New York: ACLU, 2004).

63. Stanley, *The Surveillance-Industrial Complex*, 25.

64. *"Amnesia,"* in *We Know You Are Watching*, 25.

Chapter 5

Epigraph: Charlie Quimby, *Monument Road* (New York: Torrey House Press, 2013), 83.

1. http://www.sciencemuseum.org.uk/broughttolife/people/cosmasdamian.aspx (accessed February 24, 2017).

2. http://www.sscosmandamiano.com/history.html (accessed February 24, 2017).

3. Graham Sewell and James R. Baker, "Neither Good nor Bad, but Dangerous: Surveillance as an Ethical Paradox," in Hier and Greenberg, *The Surveillance Studies Reader*, 360.

4. Sewell and Baker, "Neither Good nor Bad," 361.

5. Shital Pravinchandra, "The Third-World Body Commodified: Manjula Padmanabhan's *Harvest*," *eSharp* 8. http://www.gla.ac.uk/media/media_41208_en.pdf (accessed June 16, 2013).

6. Vivekanand Jha, "Paid Transplants in India: The Grim Reality," *Nephrol Dial Transplant* 19.3 (2004): 541.

7. Nancy Scheper-Hughes, "The Ends of the Body: Commodity Fetishism and the Global Traffic in Organs," *SAIS Review* 22.1 (2002): 70.

8. Jha, "Paid Transplants in India," 542.

9. See, for example, Urie Bronfenbrenner, "Freudian Theories of Identification and Their Derivatives," *Child Development* 31.1 (1960): 20, for a more detailed discussion of Anna Freud, Bruno Bettelheim, and the concept of identifying with the aggressor.

10. Manjula Padmanabhan, *Harvest* (London: Aurora Metro Press, 2003), 47.

11. Padmanabhan, *Harvest*, 85.

12. Viktor Mayer-Schönberger and Kenneth Cukier, *Big Data: A Revolution That Will Transform How We Live, Work and Think* (New York: John Murray, 2013), 5.

13. Mayer-Schönberger and Cukier, *Big Data*, 2.

14. Mayer-Schönberger and Cukier, *Big Data*, 100.

15. Alice E. Marwick, "How Your Data Are Being Deeply Mined," *New York Times Review of Books*, January 9, 2014. www.nybooks.com/articles/archives/2014/jan/09/how-your-data-are-being-deeply-mined/?pagination=false&printpage=true (accessed May 19, 2014).

16. Mayer-Schönberger and Cukier, *Big Data*, 56.

17. Jill A. Fisher and Torin Monahan, "Tracking the Social Dimensions of RFID Systems in Hospitals," *International Journal of Medical Informatics* 77 (2008): 178.

18. Fisher and Monahan, "Tracking the Social Dimensions," 176.

19. Fisher and Monahan, "Tracking the Social Dimensions," 179.

20. See Cheryl L. Brown, "Health-Care Data Protection and Biometric Authentication Policies: Comparative Culture and Technology Acceptance in China and in the United States," *Review of Policy Research* 29.1 (2012): 141–159.

21. Emilio Mordini and Sonia Massari, "Body, Biometrics and Identity," *Bioethics* 22.9 (2008): 490.

22. Mordini and Massari, "Body, Biometrics and Identity," 491–492.

23. Mordini and Massari, "Body, Biometrics and Identity," 492.

24. Daniel J. Solove, "The Digital Person and the Future of Privacy," in *Privacy and Technologies of Identity: A Cross-Disciplinary Conversation*, ed. Katherine J. Strandburg (New York: Springer, 2006), 4.

25. http://www.wilanddirect.com/catalog-and-retail-solutions (accessed April 24, 2017).

26. Helen Gilbert, "Manjula Padmanabhan's *Harvest*: Global Technoscapes and the International Trade in Human Body Organs," *Contemporary Theatre Review* 16.1 (2006): 29.

27. Lyon, *Surveillance Studies*, 197.

28. Beth A. Kattelman, "Spectres and Spectators: The Poly-Technologies of the Pepper's Ghost Illusion," in *Theatre, Performance and Analogue Technology* , ed. Kara Reilly (London: Palgrave Macmillan, 2013), 198.

29. Helen Groth, "Reading Victorian Illusions: Dickens's Haunted Man and Dr. Pepper's 'Ghost,'" *Victorian Studies* 50.1 (2007): 55.

30. Don Slater and George Ritzer, "Interview with Ulrich Beck," *Journal of Consumer Culture* 1.2 (2001): 262.

31. See R. A. Clark, "Information Technology and Dataveillance," *Communications of the ACM* 31.5 (1988): 499.

32. Solove, "Digital Person," 4.

33. Kathy A. Gates, "Biometrics and Post-9/11 Technostalgia," *Social Text* 83 (2005): 47.

34. Marwick, "How Your Data Are Being Deeply Mined."

35. Ishwar K. Sethi, "Biometrics: Overview and Applications," in Strandburg, *Privacy and Technologies of Identity*, 117.

36. Gates, "Biometrics and Post-9/11 Technostalgia," 50.

37. Although Lyon has written widely on the concept of "social sorting," one of the concise articulations of this notion can be found in David Lyon, "Surveillance, Security and Social Sorting: Emerging Research Priorities," *International Criminal Justice Review* 17.3 (2007): 161–170.

38. Cameron, "CCTV and (In)dividuation," 141.

39. Gates, "Biometrics and Post-9/11 Technostalgia," 36.

40. Cameron, "CCTV and (In)dividuation," 141.

41. Since 2008, Booz Allen Hamilton has focused all of its attention on a single client: the U.S. government. Apparently, this has not hurt the company. As Drake Bennett and Michael Riley have noted: "In the fiscal year ended in March 2013, Booz Allen Hamilton reported $5.76 billion in revenue, 99 percent of which came from government contracts, and $219 million in net income." See "The World's Most Profitable Spy Organization," BloombergBusinessweek, June 21, 2013. http://www.businessweek.com/articles/2013-06-20/booz-allen-the-worlds-most-profitable-spy-organization (accessed April 24, 2017).

42. Padmanabhan, *Harvest*, 91, 90.

43. Padmanabhan, *Harvest*, 91, 90.

Chapter 6

Epigraph: Badiou, *Rhapsody for the Theatre*, 117–118.

1. Georg Patzer, "Keksdose Internet," October 12, 2014. http://www.nachtkritik.de/index.php?option=com_content&view=article&id=10110:2014-10-13-06-10-03&catid=85:badisches-staatstheater&Itemid=100190 (accessed January 29, 2016).

2. http://www.staatstheater.karlsruhe.de/programm/presse/1938/ (accessed January 30, 2015).

3. Nick Kaye, "Screening Presence: The Builders Association and dbox, SUPER VISION (2005)," *Contemporary Theatre Review* 17.4 (2007): 557.

4. Maurya Wickstrom, "Data Bodies and the Awesome Apparatus of Technology," *PAJ* 28.2 (2006): 97.

5. Wickstrom, "Data Bodies," 97.

6. Andy Lavender, "The Builders Association—Super Vision (2005)—Digital dataflow and the Synthesis of Everything," in *Making Contemporary Theatre: International Rehearsal Processes*, ed. Jen Harvey and Andy Lavender (Manchester: Manchester University Press, 2010).

7. Morrison, *Discipline and Desire*, 272.

8. Karl Marx, *The 18th Brumaire of Louis Bonaparte* (New York: International Publishers, 2015), 15.

9. Crouch, *Post-democracy*, vii.

10. Crouch, *Post-democracy*, 41.

11. Crouch, *Post-democracy*, 41.

12. Crouch, *Post-democracy*, 89.

13. Metahaven (Vinca Kruk and Daniel van der Velden), *Black Transparency: The Right to Know in the Age of Mass Surveillance* (Berlin: Sternberg Press, 2015), 35.

14. Crouch, *Post-democracy*, 14.

15. Crouch, *Post-democracy*, 12.

16. Lukasz Chrobok, "Episode III—Roadshow 2006." http://www.lukaszchrobok.de/works/mister-security/episode-iii/ (accessed June 16, 2016).

17. Lukazs Chrobok and Christoph Faulhaber, *Mr. Security: To Serve and Observe* (Frankfurt am Main: Revolver Verlag, 2007), 40.

18. Chrobok and Faulhaber, *Mr. Security*, 39.

19. Oliver Zybok, exhibition catalog, *Mr. Security*, 2.

20. Chrobok and Faulhaber, *Mr. Security*, 55.

21. Chrobok and Faulhaber, *Mr. Security*, 9.

22. Artworks Copyright Law cited in Chrobok and Faulhaber, *Mr. Security*, 15.

23. Marilyn Frankenstein, "Studying Culture Jamming to Inspire Student Activism," *Radical Teacher* 89 (2010): 33.

24. Cited in Frankenstein, "Studying Culture Jamming," 33.

25. Schechner, *Performance Studies*, 119–122.

26. Chrobok and Faulhaber, *Mr. Security*, 52.

27. Letter from Claire Montgomery [executive director of Location One] to Dr. Daniele Perrier [director of Künstlerhaus Schloß Balmoral], September 16, 2008. Published in Christoph Faulhaber, *Ich: Wie Es Wirklich War—Ein Bericht Aus New York* (Berlin: Revolver Publishing, 2011), n.p.

28. Badiou, *Rhapsody for the Theatre*, 117.

29. Badiou, *Rhapsody for the Theatre*, 117.

30. Badiou, *Rhapsody for the Theatre*, 117.

31. Taitu Heron, "Globalization, Neoliberalism and the Exercise of Human Agency," *International Journal of Politics, Culture and Society* 20.1 (2008): 89.

32. Crouch, *Post-democracy*, 22–23.

33. Jean Baudrillard, *The Spirit of Terrorism and Requiem for the Twin Towers*, trans. Chris Turner (London: Verso, 2002), 15.

34. Baudrillard, *The Spirit of Terrorism*, 8–9.

35. Baudrillard, *The Spirit of Terrorism*, 57.

36. Baudrillard, *The Spirit of Terrorism*, 19.

37. Creative Art Ensemble, "Nomadic Power and Civic Disobedience," in *The Electronic Disturbance* (New York: Automedia, 1994), 15.

38. Creative Art Ensemble, "Nomadic Power," 23.

39. Creative Art Ensemble, "Nomadic Power," 12.

40. Baudrillard, *The Spirit of Terrorism*, 19.

41. Brian Holmes, "Three Keys and No Exit: A Brief Introduction to Critical Art Ensemble," in Critical Art Ensemble, *Disturbances* (London: Four Corner Books, 2012), 13.

42. Bruce Sterling, *The Epic Struggle of the Internet of Things* (Moscow: Strelka Press, 2014), Kindle ed., Kindle location 63.

43. Jon McKenzie, "!nt3rh4ckt!v!ty," *Style* 33.2 (1999): 286.

44. Keller Easterling, "An Internet of Things." http://www.e-flux.com/journal/an-internet-of-things/ (accessed March 26, 2016).

45. Easterling, "An Internet of Things."

46. Bruce Sterling, "The Internet of Things Is an Oligarch Power Play." http://www.wired.co.uk/magazine/archive/2015/02/ideas-bank/the-internet-of-things-is-an-oligarch-power-play (accessed March 26, 2016).

47. Sterling, *Epic Struggle*, Kindle locations 58–59.

48. Sterling, *Epic Struggle*, Kindle locations 9–10.

49. Metahaven, *Black Transparency*, 132.

50. Metahaven, *Black Transparency*, 3.

51. Metahaven, *Black Transparency*, 132.

52. McKenzie, "!nt3rh4ckt!v!ty," 286.

53. McKenzie, "!nt3rh4ckt!v!ty," 292.

54. Catherine Bernard, "Bodies and Digital Utopia," *Art Journal* 59.4 (2000): 30.

55. McKenzie, "!nt3rh4ckt!v!ty," 292.

56. Anna Nemtsova, "Russian Artist Sets Fire to Spy Headquarters," *Daily Beast*, November 11, 2015. http://www.thedailybeast.com/articles/2015/11/09/russian-artist-sets-fire-to-spy-headquarters.html (accessed July 15, 2016).

57. Jill Lane and Ricardo Dominguez, "Digital Zapatistas," *TDR* 47.2 (2003): 135.

58. Molly Sauter, "LOIC Will Tear Us Apart": The Impact of Tool Design and Media Portrayals in the Success of Activist DDOS Attacks," *American Behavioral Scientist* 57.7 (2013): 992.

59. Sauter, "LOIC," 992.

60. Lane and Dominguez, "Digital Zapatistas," 139.

61. Gabrielle Coleman, *Hacker, Hoaxer, Whistleblower, Spy: The Many Faces of Anonymous* (New York: Verso, 2014. 196.

62. Lane and Dominguez, "Digital Zapatistas," 135.

63. Lucy Ash, "Why Russians Back Anti-police Rage," *BBC News*, March 6, 2012. http://www.bbc.com/news/world-europe-11829793 (accessed August 9, 2016).

64. Leonid Bershidsky, "The Russian Art That Won't Be Sold at Christie's," *Bloomberg View*, November 11, 2015. http://www.bloomberg.com/view/articles/2015-11-11/the-russian-art-that-won-t-be-sold-at-christie-s (accessed July 14, 2016).

65. "Russian Artist Jailed for 30 Days before Trial after Setting Fire to Security Service HQ," *Guardian*, November 10, 2015. https://www.theguardian.com/world/2015/nov/11/russian-artist-pavlensky-jailed-30-days-pending-trial-setting-fire-security-service-hq (accessed July 12, 2016).

66. Nemtsova, "Russian Artist Sets Fire," 5.

67. Lorena Muñoz-Alonso, "Criminal Charges against Russian Dissident Artist Pyotr Pavlensky Suddenly Changed From Vandalism to "Damaging a Cultural Heritage Site," *Artnet News*, March 30, 2016. https://news.artnet.com/art-world/pyotr-pavlensky-criminal-charges-changed-462155 (accessed July 12, 2016).

68. Amy Lin, "Performance Artist Pyotr Pavlensky Trial Stirs Up Even More Controversy," *Widewalls*, no date. http://www.widewalls.ch/pyotr-pavlensky-trial-controversy/ (accessed July 16, 2016).

69. Peter Hobson, "Radical Russian Artist Pyotr Pavlensky Walks Free after Verdict," *Moscow Times*, June 8, 2016. http://www.themoscowtimes.com/news/article/radical-russian-artist-pyotr-pavlensky-walks-free-after-verdict/571454.html (accessed July 13, 2016).

70. Pyotr Pavlensky, "Russian Artist Petr Pavlensky: 'I Saw the Machine from the Inside,'" interview by Dmitry Volchek, *Guardian*, June 11, 2016. https://www.theguardian.com/world/2016/jun/11/russian-artist-petr-pavlensky-i-saw-the-machine-from-the-inside (accessed July 16, 2016).

71. "Russian Performance Artist Pavlensky Says FSB Damaged Cultural Heritage," *Moscow Times*, June 7, 2016. http://www.themoscowtimes.com/news/article/russian-performance-artist-pavlensky-says-fsb-damaged-cultural-heritage/571453.html (accessed July 13, 2016).

72. Noah Sneider, "Body Politics," *The Economist* 1843, June–July 2016. https://www.1843magazine.com/features/body-politics (accessed July 16, 2016).

73. Hili Perlson, "Russia Cancels Top Art Prize after Dissident Artist Pyotr Pavlensky Is Nominated," *Artnet News*, February 17, 2016. https://news.artnet.com/art-world/russia-rejects-pyotr-pavlensky-art-prize-428654 (accessed July 12, 2016).

74. Maryam Omidi, "Russian Performance Artist Arrested for Ukraine Solidarity Protest," *Calvert Journal*, February 24, 2014. http://calvertjournal.

com/news/show/2100/russian-artist-arrested-ukraine-solidarity-protest-pyotr-pavlensky (accessed July 14, 2016).

75. Hili Perlson, "Radical Dissident Artist Pyotr Pavlensky Refuses Court's Proposal to End His Trial," *Artnet News*, April 27, 2016. https://news.artnet.com/art-world/pavlensky-russian-dissident-artist-refuses-end-trial-482762 (accessed July 14, 2016).

76. Marc Bennetts, "Acts of Resistance: Pyotr Pavlensky on Performance Art as Protest," *Calvert Journal*, December 1, 2014. https://calvertjournal.com/articles/show/3373/pavlensky-performance-art-protest (accessed July 14, 2016).

77. Dmitry Volchek, "Russian Protest Artist Pavlensky Alleges Police Brutality," *Radio Free Europe—Radio Liberty*, May 18, 2016. http://www.rferl.org/content/russia-artist-pavlensky-complains-police-brutality/27743045.html (accessed July 15, 2016).

78. "Law and Disorder: Russian Protest Artist Calls Prostitutes as Witnesses," *Art Newspaper*, April 28, 2016. http://theartnewspaper.com/news/in-the-frame/law-and-disorder-russian-protest-artist-calls-prostitutes-as-witnesses/ (accessed July 15, 2016).

79. Perlson, "Radical Dissident Artist."

80. Perlson, "Radical Dissident Artist."

81. Pyot Pavlensky, "Interrogation: A Play in Three Acts," *Snob*, July 29, 2014. https://snob.ru/selected/entry/77648 (accessed August 3, 2016).

82. Sneider, "Body Politics."

83. Steve Mann, "Sousveillance: Inverse Surveillance in Multimedia Imaging," in *International Multimedia Conference Proceedings of the 12th ACM International Conference*, New York, October 10–16, 2004, 627.

84. Bennetts, "Acts of Resistance."

85. As recently as 2013, for example, Mann has argued that sousveillance can "serve as a potentially less-biased and more neutral feedback mechanism than police-owned surveillance-only media." See Steve Mann, "Veilance and Reciprocal Transparency: Surveillance versus sousveillance, AR Glass, Lifeglogging, and Wearable Computing," IEEE International Symposium on Technology and Society (ISTAS), "Social Implications of Wearable Computing and Augmediated Reality in Everyday Life," June 2013, 5.

86. Bennetts, "Acts of Resistance."

87. Sneider, "Body Politics."

88. Sneider, "Body Politics."

89. Pavlensky, "Russian Artist Petr Pavlensky."

90. Pavlensky, "Russian Artist Petr Pavlensky."

91. Susan Buck-Morss, *Thinking Past Terror: Islamism and Critical Theory on the Left* (New York: Verso, 2003), 68.

92. Buck-Morss, *Thinking Past Terror*, 69.

93. Online Etymological Dictionary, http://www.etymonline.com/index.php?term=pyrotechnic (accessed July 29, 2016).

94. Alexey Plutser-Sarno, "Towering Penis Graffiti Startles St. Petersburg Residents," *The Observers*, June 22, 2010. http://observers.france24.com/en/20100622-giant-penis-towers-over-st-petersburg-skyline-russia-liteiny-bridge-art-voina (accessed July 27, 2016).

95. Sneider, "Body Politics."

96. Sergey Gulekin, "Red Square Riots: Performance Art in the Centre of Moscow," *Calvert Journal*, June 3, 2015. http://calvertjournal.com/articles/show/4190/performance-art-in-red-square-pussy-riot- (accessed July 25, 2016).

97. Gulekin, "Red Square Riots."

98. Sneider, "Body Politics."

99. Ekow Eshun, Maryam Omidi, Jamie Rann, and Igor Zinatulin, "The Naked Truth: The Art World Reacts to Pyotr Pavlensky's Red Square Protest," *Calvert Journal*, November 14, 2016. https://calvertjournal.com/articles/show/1768/pyotr-pavlensky-russian-artist-nails-red-square (accessed July 13, 2016).

100. Eshun et al., "The Naked Truth."

101. Eshun et al., "The Naked Truth."

102. Sneider, "Body Politics."

103. Sneider, "Body Politics."

104. Elise Morton, "Russian Art Activist Pyotr Pavlensky Claims Assault by Police," *Calvert Journal*, May 18, 2016. http://calvertjournal.com/news/show/6038/russian-art-activist-pyotr-pavlensky-claims-assault-by-police (accessed August 13, 2016).

105. Torin Monahan, "Counter-surveillance as Political Intervention?," *Social Semiotics* 16.4 (2006): 516–517.

106. Monahan, "Counter-surveillance," 523.

Index